Cartographic Methods

Cartographic Methods

G. R. P. Lawrence

Methuen

LONDON AND NEW YORK

still current BBIP 3/87 √

First published in 1971
Second edition published in 1979 by
Methuen & Co. Ltd
11 New Fetter Lane, London EC4P 4EE
Published in the USA by
Methuen & Co.
in association with Methuen, Inc.
733 Third Avenue, New York, NY 10017
© 1971 and 1979 G. R. P. Lawrence

Filmset by
Northumberland Press Ltd, Gateshead, Tyne and Wear
Printed in Great Britain by
The University Press, Cambridge

ISBN 0 416 71640 7   (hardback edition)
ISBN 0 416 71650 4   (paperback edition)

British Library Cataloguing in Publication Data

Lawrence, George Richard Peter
    Cartographic methods. – 2nd ed.
    1. Cartography
    I. Title
    526   GA105.3   79–40733

    ISBN 0–416–71640–7
    ISBN 0–416–71650–4 Pbk

For Catherine

Contents

Text figures, map extracts and tables

TABLES

Preface to the first edition

This book sets out to provide a general survey of those aspects of cartography of relevance and importance to the geographer. Just as the study of geography embraces a variety of different outlooks, cartography involves its practitioners in a number of subjects, ranging from the intensely practical work of the map-printer to the artistic demands of map design. The geographer, in handling and making maps for his own purposes of analysis and also of data presentation, provides a further meaning to the term cartographer. The entire field of cartography is therefore large and a complete handbook would run into several volumes. I have tried to select a number of pertinent detailed aspects in a short space, and to provide a few signposts to the more detailed treatment of the subject, particularly by quoting some of the more accessible references in other books and papers.

Although some of the illustrations in this book are taken from published material and are appropriately acknowledged, the great majority of the diagrams have been drawn by my wife to whom I am greatly indebted, not only for draughting but also for much help and encouragement in preparing this book. Thanks are also due to Miss Pat Aylott, departmental photographer, Geography Department, University of London King's College for help in preparing photographic copies. Much other assistance was also provided by the cartographers, technical and secretarial staff in the same department and is also gratefully acknowledged. Finally, I am greatly indebted to both Professor J. C. Pugh and Dr W. B. Morgan as well as the publishers for their editorial advice and assistance.

1971

King's College
London

Preface to the second edition

In preparing a second edition of this book it has been possible to introduce various changes in cartography that have occurred since the original was prepared. In terms of British cartography this period has been marked by the introduction of metric units and standards and by the replacement of the well-known 1 inch to the mile map by the 1/50,000 series.

The subject of cartography has been characterized in the same period by a greater awareness of its importance as a means of communication, and this aspect has also received brief coverage. Some of the material previously used in the appendices has been incorporated in the main text, which has generally been up-dated.

1979

King's College
London

Acknowledgements

The author and the publishers would like to thank the following
for permission to reproduce copyright material:

The lower of the maps in the frontispiece is reproduced with the permission
   of the Federal Office of Topology, Switzerland from June 18 1979.
The Controller of Her Majesty's Stationery Office (Crown Copyright
   reserved) for Figures 1.3a, 2.1, 3.7b, 4.2b, 4.4, 9.3 and material used
   in Appendix 1.
The US Geological Survey for Figures 1.3b and 3.7a.
The Institut Geographique National, Paris for Figure 1.3c.
Niedersächsisches Landesverwaltungsamt, Bavaria for Figure 1.3d.
Professor E. H. Brown for Figure 1.4d.
Professor E. Imhof for Figure 2.6.
The Controller of Her Majesty's Stationery Office and the Hydrographer of
   the Navy for Figures 2.8 and 2.9.
Pergamon Press Ltd for Figure 3.2.
Dr M. C. Storrie and Dr C. I. Jackson for Figure 3.4.
Professor R. H. Osborne for Figure 3.5.
Topografische Dienst, Delft for Figure 4.1.
Fairey Surveys Ltd for Figure 4.2.
Hodder & Stoughton Educational for Figure 4.5.
The National Geographic Society, Washington, D.C., for Figure 6.1.
The Institute of British Geographers for Figures 9.9 and 10.6.
Geographical Publications Ltd for Figure 10.4.
George Philip and Son Ltd for Figure 11.3.
Graphic Systems International Ltd for Figure 12.10.

Introduction

The geographer's field of study is sometimes precisely, sometimes loosely defined. The subject matter involved deals not only with the surface features of the earth but also with the interrelationships between man and his environment. The presentation of these facts and their analysis in depth often involve specialization in disciplines beyond the realm of geography, but the geographer is usually recognized as the expert when spatial patterns and relationships are under discussion. Although these patterns may be represented in the abstract way made possible by mathematical and statistical studies it is often simpler to perceive them in a pictorial manner. Such is the field of mapping to the geographer – both as a source of information and as a method of presentation. Even so, the whole range of cartography covers a wide field. It is convenient to divide this field into those sections which deal with the presentation of information and those which consider methods of studying mapped data. Obviously the former is coupled with a study of one of the basic sources of geographical data – the topographic map – and here one enters the field of surveying as well as cartography. Techniques of analysis depend to a certain extent on the interest within the subject – thus the geomorphologist's approach may well differ from that of the historical or the urban geographer. Nevertheless there are many points of similarity in these analyses as all are concerned with distributions in geographical space.

The first section of this book considers the basic function of any map or cartographic system – the need for scale representation and the presentation of information according to cartographic rules and conventions. How these rules are applied in various published maps is considered in the second section, which examines British and overseas map series, as well as atlases. Cartographic methods of study involving some modern quantitative techniques make up the third section and the concluding section briefly discusses a number of ways in which data may be presented in cartographic form for research or publication.

1

The basic framework: scales and grids

A map may be defined as a representation to scale of the features of the surface of the earth. These features present a number of problems to the map maker since they include both visible items such as hills, valleys, buildings and roads and also important but invisible features such as boundaries, place-names and frontiers.

Many of the visible features can be portrayed well in a photograph, and such a view, almost to scale, can be given by a vertical aerial photograph or similar 'remote sensor'. This latter means of surveying the earth's surface can detect an even wider range of information because such sensors respond to a wider band of the electro-magnetic spectrum. Nevertheless, all this information can most easily be conveyed to the user in the form of a map.

The basic differences between the map and the aerial photograph or the image obtained from other remote sensing are that elements of interpretation and selection have been undertaken by the map compiler. Much map data are presented in the form of point or line information; the point data may further be simplified (or codified) as a symbol while the line may either depict a routeway such as a track or demarcate the edge of some distribution. Arbitrary information that does not exist on the ground, for example place-names, administrative boundaries and spot heights, can also be shown on a map. The aerial photograph shows *all* the visible details of the land surface, presenting the information to the viewer by a series of tones which may be in colour but are more likely to be of various greys as well as in black and white.

A single aerial photograph cannot tell us anything about the three-dimensional aspect of the landscape, the relief of the land and its height, except perhaps by way of shadows. This can be overcome by using stereoscopic pairs of aerial photographs, from which the user can obtain a 'space model' or mental picture in three dimensions. The map has to use other conventions to present height information (see Chapter 2).

Both the vertical aerial photograph and the map are scale representations of the earth's surface. By this is meant that accurate measurements made

Figure 1.1   Vertical air photograph of part of the inner urban area of Christchurch, New Zealand, together with monochrome enlargement from the 1/63,360 topographic map of the same area. This area is very nearly level, so height distortions due to local relief are negligible. The greater detail of the black and white air photograph can easily be appreciated. The scale of the air photograph is approximately 1/20,000 and the map 1/50,000.

on the map are proportional to their true values on the surface of the ground, or that every point is correctly positioned relative to all other points. Measurements from aerial photographs are less accurate than those on maps because of various distortions; but nearly all maps are prepared as scale representations on a plane surface, normally the surface representing sea-level. On small scale maps covering extensive areas of the earth's surface, where the problem of curvature of that surface arises, it is necessary to consider the map projection that has been used and the properties this possesses. For the purposes of the discussions in this volume it will be assumed that the distortions resulting from the curved surfaces of the earth have been removed by the map maker's choice of projection or that the scale of the map is sufficiently great for these errors to be disregarded. A summary of the main characteristics of map projections will be found on pages 141–6.

REPRESENTATION OF SCALE

Scale is defined as the relationship between a distance measured on the map and the true distance on the ground. The natural scale is usually expressed as a ratio or 'Representative Fraction' and metric relationships are increasingly used, e.g. 1/10,000 or 1/50,000. Nevertheless, many maps in the English-speaking world still include Imperial units, e.g. '1 inch to the mile', and these give rise to such ratios as 1/63,360, there being 63,360 inches in 1 mile. Other unusual scales may be encountered in old maps or maps prepared to be printed on a given sheet size, such as older nautical charts or atlas maps. Divergences from the intended natural scale may occur in the production of aerial photographs by reason of problems of maintaining flying height or large differences in land height. If no scale value is given for a map, either by a fraction, printed scale bar or statement in words, it must be calculated by comparing a measured length on the map with its known length on the ground. In the case of aerial photographs the flying height (H) and the focal length (f) of the camera used may be quoted in the marginal

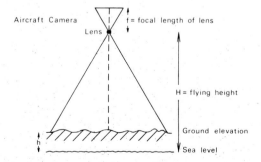

Figure 1.2 Relationship between aircraft flying height, focal length of camera lens and scale of an air photograph

information. The scale can then be calculated from the expression $\dfrac{f}{H-h}$ where h is the height of the ground above mean sea-level (Figure 1.2). Care must be taken in all these calculations to ensure that comparable units are used (inches, feet, miles, or metres, kilometres, etc.). For instance:

a 1 cm on the map representing 1 km on the ground gives a representative fraction of 1/100,000.

b An aerial photograph taken from a height of 5,000 ft by a camera of focal length 6 ins will have a scale of 1/10,000 if the area photographed is at sea-level, or

$$\frac{6 \text{ ins}}{(5,000 \text{ ft}) - (1,000 \text{ ft})} = \frac{1}{8,000}$$

if the area photographed is 1,000 ft above sea-level.

c If 0·5 ins on the map represents 2 miles on the ground, the scale is calculated as 0·5 represents 2 × 63,360 ins or 1/253,440, approximately 1/250,000 (Quarter Million).

Map scales are extremely varied. In general, maps at a very large scale, e.g. 1/500 or 1/1,250 are referred to as *plans* and show detail precisely and with no exaggeration and little variation from the ground shapes they portray. The map projection used by such plans will be barely obvious to the user. Small scale maps such as those of countries or whole continents will require to be presented on a map projection graticule and are best referred to as *atlas* maps. It is generally found that simplifications of ground detail begins between the scales of 1/5,000 and 1/10,000 and, once it has been found necessary to introduce generalizations of shape or to omit items of detail, the cartographer's work becomes subjective. The degree of subjectivity may be reduced by a form of shorthand in which particular objects or features are replaced by symbols. Frequently, however, it is necessary to suppress information in order that chosen symbols, or even the necessary description, naming or lettering, may be placed on the map.

In connection with map scale the words 'small', 'medium' and 'large' are frequently used. There is little general agreement as to where the dividing lines come. Recent usage by British organizations has been as shown in the table below:

	Ordnance Survey	Directorate of Military Surveys	Directorate of Overseas Surveys
Large scale	1/10,000	1/75,000	1/25,000
Medium scale	—	1/75,000 to 1/600,000	1/25,000 to 1/125,000
Small scale	1/25,000	1/600,000	1/125,000

It may sometimes be helpful to recognize a fourth category of 'atlas scale' for maps at much smaller scales, e.g. 1/1,000,000 (1/1 million) or smaller.

This discussion of the significance of the scale of a map must also emphasize the fact that any line or symbol drawn on a map has a definite size. Even the finest printed line will represent a strip of land and thus reduce the amount of space available on the map for other information.

Figure 1.3 Scale bars taken from topographic maps: the German extract also includes beneath the scale bar part of a slope scale by which terrain angles can be calculated from measurement of contour spacings.

    a Ordnance Survey 1/50,000
    b United States 1/62,500
    c French 1/50,000
    d German 1/50,000

For example, a line 0·1 mm wide will cover a width of 5 m on a 1/50,000 map. The aerial photograph shows many lines by the boundary between different tones or colours and often in these cases no measurable thickness can be resolved. On the printed map, however, the boundaries between land parcels, road edges, etc. are shown as printed lines. It is customary to regard these lines as having no thickness but errors may arise, for example, in measuring distributions and areas on such maps.

The problem of scale is encountered at some stage in all geographical studies. One possible framework that can link all variations in size from the entire earth's surface down to the individual land parcel is the *G-scale* (Haggett 1965). Using this scale and a map format of 1,000 sq cm in area the following table indicates variations in size of land surface covered by maps at different scales (see also Figure 1.3). Thus a 1/10,000 map of this size covers 10 sq km whilst a 1/100,000 map will include 1,000 sq km. The respective G-scale values for these two cases are 7·7074 and 5·7074.

G-scale value for map sheet of 1,000 sq cms format

Large scale plans (e.g. 1/2,500)	G-scale 8·5033
Medium scale maps (e.g. 1/25,000)	G-scale 6·5033
Small scale maps (e.g. 1/250,000)	G-scale 4·5033
Atlas maps (e.g. 1/2,500,000)	G-scale 2·5033

It is obvious from all the foregoing comments that a small denominator in the representative fraction characterizes the large scale map whereas a large denominator means a small scale. Large scale map sheets cover only a small area but small scale maps are needed to show a large area on a single sheet of paper. The scale quoted on many atlas maps may only apply to certain lines (of latitude or longitude) and in some cases is really the scale of the appropriate sphere from which the projection has been made.

Figure 1.4   Geographical scale, map scale and area: diagram based on the concept of 'G-scale' (Haggett 1965)

MAP TYPES

Maps are of very varied size and appearance as well as of differing scales but in general it is possible to distinguish between two major categories of maps. On the one hand there are maps that depict as much of the surface features of a particular area as is possible within the limitations of the scale. Such maps are called *topographic* maps and familiar examples of these are the 1/25,000 and 1/50,000 national map series produced by many countries.

On the other hand many maps are designed for specialized purposes. They may be drawn to indicate certain distributions and/or to supplement the basic data of the topographic map. In many cases all or part of the topography will be shown as a base map with the additional information drawn or overprinted upon this framework. Examples of such cases are maps of geological outcrops and land utilization, and soil and vegetation maps. Again, the particular purpose of the map may best be served with only the minimum framework of reference points. The common factor in all these maps is that they have a single theme of study and they are therefore grouped together as *thematic* maps.

It is clear from the above that the so-called topographic maps are themselves thematic, in that they have been prepared for a specific purpose. This purpose is to present an overall view of the topography of the particular area for a variety of users: in many cases the needs of a particular user – e.g. the military authorities – may have been paramount. The picture of a countryside that they portray is limited in various ways. Although a policy of 'continuous revision' may be adopted, as with most current Ordnance Survey productions (see pp. 52–5), the topographic map presents a 'synoptic' picture of the land surface at the time of the survey. For some geographical purposes this may be adequate; for example, a series of maps of different dates will be of value to any study of the historical development of a region. This *static* framework of the topographic map is generally adequate as a basis for presenting the results of a geographical study, but in many branches of modern geography an emphasis is placed on dynamic characteristics, such as changes, processes, development and functional relationships or the potential for these.

Board (1967, in Chorley and Haggett) distinguishes between *single-purpose* and *multi-purpose* maps. Although thematic maps can belong to the latter category they are more usually single-purpose in character whilst topographic maps are multi-purpose. For various study purposes geographers, economists, planners and so forth need to prepare and/or make use of thematic maps and it is useful further to subdivide this category. The *diagrammatic* presentation of information can be based on its real spatial relationships (i.e. utilizing natural positions on the earth's surface) or according to relationships obtained for instance from theoretical considerations, non-Euclidean geometries, model-building theories and the

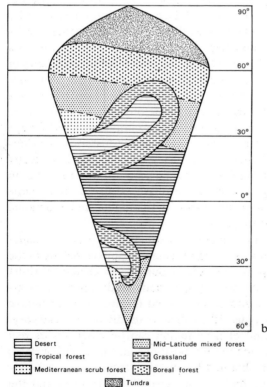

Figure 1.5 Cartogram examples

  a Line diagram of London suburban railways (a topographical map or *route* cartogram)

  b World climatic regions (a diagrammatic *regional* cartogram)

Desert

Tropical forest

Mediterranean scrub forest

Mid-Latitude mixed forest

Grassland

Boreal forest

Tundra

c

c Urban hierarchies in
the Netherlands
(cartogram of a stage
in urban analysis)
d Contours of the
Palaeozoic platform
(after E. H. Brown 1964)
(a *theoretical* cartogram)

d

like. Such maps are frequently referred to as *cartograms* and they have also been called topological maps (Cole and King 1968). The selective presentation of results of study or of material for purposes of further study in diagram form might be separately classed as *analytic* maps. Examples of these categories are:

> *Cartograms*: British Rail, London Transport route maps
> *Diagrammatic*: Urban hierarchies, functional regions
> *Analytic*: Generalized contours.

REFERENCE SYSTEMS

In all mapping it is necessary to be able to locate individual points with some degree of precision and, for the map maker, the precise location of detail will be one of the major stages in compilation from field data. The printed map will eventually be consulted by the map-user who will wish to extract information from it, or to locate further data on it. The plane surface of the map sheet will thus need a reference framework by which points can

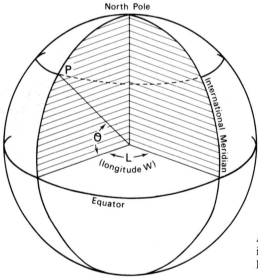

Figure 1.6   The spherical earth indicating the definition of latitude and longitude

Figure 1.7   The relationship between the earth's graticule (lines of latitude and longitude) and an arbitrary grid, in this case the National Grid of the Ordnance Survey of Great Britain: only the southern 1,000 km of the grid are indicated; the northern continuation to include the Orkneys and Shetland is not shown. Note the coincidence of the 2°W meridian and the 400 km grid line, also the position of the false origin so that all references are positive. The true origin of the orthomorphic Transverse Mercator projection on which the grid is based is at 49°N and 2°W.

be sited or found. On the surface of the earth an arbitrary and well-known system of references is used – lines of latitude and longitude, which form a graticule (Figure 1.6). These are generally used on map series but they suffer from a major defect: the earth's surface is curved and, although the lines of latitude and longitude are straight and intersect at right angles on this surface, they cannot do so on a flat sheet and also preserve the angular relationships desired by the map maker. However, two-dimensional reference systems used to locate points are normally based on a rectilinear grid and measurements in units of distance (kilometres, miles, etc.) to the nearest grid line can be made correctly and easily. Measurements of latitude or longitude to graticule lines require greater care as, for example, the lengths along the parallels decrease as one moves from the Equator to the Poles.

The choice of a map grid system for referencing is made by the map maker, who will do so after consideration of problems of map projection, layout of the sheets forming a map series, desirability or otherwise of linking with map series of adjacent countries and so on. The map grid best known to British users is the National Grid Reference System used by the Ordnance Survey.

Conventionally in this form of grid, referencing points are sited and referred to in a version of *Cartesian co-ordinates* or rectangular grid co-ordinates. Thus the extreme south-western point of the system is taken as the origin and all points are related to this origin. In actual fact this origin in the National Grid is a 'false origin' in order to remove problems of negative co-ordinates. The true origin is at 2°W longitude and 49°N latitude (Figure 1.7). In order to reduce scale error at the margins of the grid the scale along the central meridan has been reduced by 0.04 per cent as the scale increases outwards from the central meridian; consequently the modification ensures only small scale errors, negative or positive, instead of errors all positive in sign but of greater percentage error away from the central meridian. These details are adequately explained in the HMSO publication *The projection for Ordnance Survey maps and plans and the National Reference System* (Ordnance Survey 1951). The National Grid System of giving references at various scales is given in Appendix 1. It should be noted that in all reference systems the distance *east* from the origin is quoted first, followed by the distance *north*, i.e. 'Eastings' before 'Northings'.

Other systems of locating points are used, or have been used on various map series. These fall mostly into the category of locating points to a specific square which is found by a combination of identifying letters and numbers in the map margins. It is less convenient than that depending on Cartesian co-ordinates and is obviously less precise. A variation on this is used by some mapping agencies in order to refer to map sheets within the series. This is well exemplified by the French national map series (Figure 1.8) in which a combination of numbers in Roman and Arabic form are used. Those in

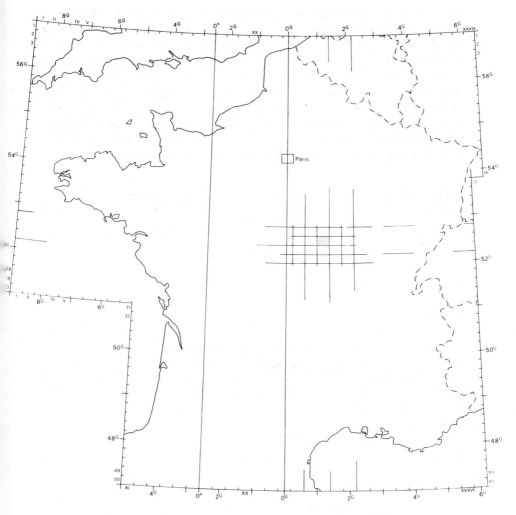

Figure 1.8 The arrangement of the topographic map sheets of France (1/50,000 scale) and their relationship to the graticule: note that the prime meridian is that of Paris and that latitude and longitude are measured in *grads* (100 grads = 90°). Map sheets are referred to by combining Roman numbers (columns) and Arabic (rows). Only a small section of the full matrix is shown here and the shaded map is XXVI–23.

Roman type give the north–south column, which is also linked with the meridians on a Conic projection graticule, while the west–east lines are shown by Arabic numerals.

The official map sheets of the United States of America are closely related to the graticule of the earth's lines of latitude and longitude with the actual sheets being based on a Conical projection. This arrangement permits the close inter-relationship between scales shown on Figure 1.9. Each map sheet at the Quarter Million scale (1/250,000) covers the same area as four sheets at twice the scale (1/125,000). The basic map scale for any area depends to a great extent on the degree of development of the region (Figure 1.10) and so thus for the state of Arizona the largest topographic map scale available for some areas is the Quarter Million. The map sheets are often referred to in terms of the graticule coverage, the largest standard scale being the 1/62,500 which is for an area of 15′ latitude by 15′ longitude, but for some areas the 7½′ quadrangle (1/24,000 or 1/31,680 scale) is also available and there are four of these to each 1/62,500 sheet.

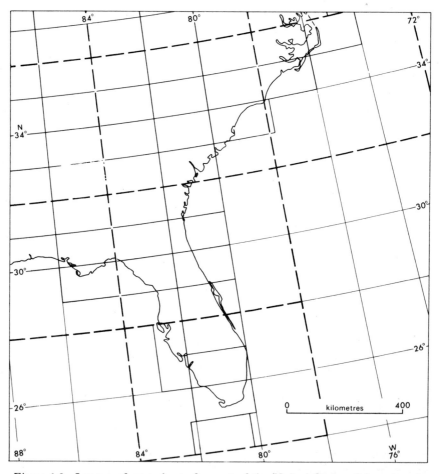

Figure 1.9   Layout of map sheets for part of the United States of America: the heavy broken lines indicate the arrangement of the large 1 to 1 million map sheets and within these the lighter rectangles are the sheet-lines of the Quarter Million map sheets (1/250,000 or approximately ¼ inch to the mile scale).

The importance of co-ordinate reference location systems is increasingly marked with the use of data banks and computer methods of storing and analysing data. A rectangular system of co-ordinates is used in most cases and it may be necessary to distinguish between the co-ordinates used on the map and those used for computations. The necessary qualities of such a location referencing system are:

  i The specified point must be unique.

  ii Relative positions must be capable of identification.

  iii The system must be capable of use with electronic data processing methods. (Maps produced by computer are considered on pp. 46–9.)

Maps published

—— River surveys

15′ quadrangles

7½′ quadrangles (Scale 1:24,000)

0   100

kilometres

Figure 1.10 Published maps for the state of Arizona at scales larger than 1/125,000: this map shows only the availability of maps at 1/62,500 (the 15′ quadrangles) and at 1/24,000 (the 7½′ quadrangles) as well as those sections of the rivers for which river surveys have been published.

2

Representation of relief, drainage and coasts

All topographic information can be seen to relate to a specific point, line or area, and many cartographic methods apply to the depiction of both the natural and artificial features of the topography. Nevertheless, some methods of representation have been developed for particular features; for instance land contouring is applied to a visible surface seen either in the field or in the space model of the stereo-viewer while other isolines represent less tangible shapes. Similarly, special symbols and mapping methods have been developed for representing hydrographic features.

This chapter will therefore consider in particular various methods of representing the natural features of the physical topography, i.e. the relief, drainage and coastlines. The representation of distributions, point and line information other than that relating to natural features will be considered in Chapter 3.

REPRESENTATION OF RELIEF

Representation of relief is probably the most difficult problem in mapping (and map reading) and stems from the basic need to present the three dimensions of the landscape on the flat, two-dimensional paper form in which the map appears (Figure 2.1).

Methods used to overcome this difficulty can be considered under two main headings: those in which a visual *impression* of relief is presented and those in which precise *values* can be set out.

However, it is first necessary to consider briefly the possible data available to the map maker. Height information may be provided by precise survey methods of levelling in which the exact height of a series of points is determined. These points may be recorded on the ground as bench marks and this elevation related to a datum or mean sea-level at some specified point. In Great Britain this datum is a level 15·588 ft below a bench mark adjacent to the Tidal Observatory at Newlyn in Cornwall. This value was calculated by taking the mean value of hourly sea-level values

Figure 2.1 Three map extracts to show relief representation methods over the centuries.
  a Extract from Saxton's map of Devon (1575): hill symbols
  b The same area on the first inch to the mile Ordanance Survey map of Dartmoor (1809): hachures
  c Monochrome print of the same area on the Dartmoor tourist edition 1 inch to the mile map (1967): spot heights, hill shading, contours and layer shading

between 1915 and 1921. A previous datum was established at Liverpool and on old large scale Ordnance Survey maps discrepancies may arise between values calculated from these two datum points. Corresponding datum values are established by all national mapping organizations and again discrepancies may be encountered when considering maps of adjacent countries, for example between the French and German topographical series in north-west Europe (Table 6.1). Surveyed height information from ground surveys based on such data is therefore available at a large number of selected points; there is no information on the way in which height variations take place between these points.

Much mapping today is carried out from aerial photography. Stereoscopic examination of pairs of air photographs provides a visual impression of the landscape in miniature and height differences can be calculated by measuring parallax displacements. Once again fairly precise values for individual points can be obtained and the cartographer has the added advantage of a view of the three-dimensional scene in the overlapping pair of air photographs. He can interpret this 'space model' visually, to obtain positions of breaks of slope, and so forth, and can plot countour lines (see below) by steering along them the floating point in the plotting machine. More advanced methods of mapping from air photographs (for example using the orthoprojector) enable contours to be plotted semi-automatically (Petrie 1962 and 1966).

QUANTITATIVE REPRESENTATION OF HEIGHT

It is thus possible to show on a map the exact value of heights of various points on the ground so represented. There would be no visual impression of landforms in such a map. These values or precise *spot heights* (for instance summit heights, depths of valleys) are, however, extremely important. The conventional representation of relief stemming from such information is that of contouring; a contour line is one on which all points are of equal vertical distance above or below a datum. It is one of the most well-known forms of *isogram*. Strictly speaking the term contour should be reserved for those lines that have been carefully surveyed on the ground as having the particular value with which they are labelled. Precise photogrammetric methods make it possible nowadays to draw contour lines of considerable accuracy, depending chiefly on the scale of the air photography involved.

It is also possible to interpolate lines of height from a close network of spot-height values and computer programs can do this automatically. The normal assumption made in such interpolation is that height values change at a steady rate between neighbouring spot heights. The resulting isogram (or isoline) cannot therefore be given the same degree of precision and should be termed a *form line*. If the contour line has been obtained by precise measurement, as by levelling and pegging out in the field followed by detailed surveying, it then merits the term *isometric* line, that is, a line on the map obtained by measurements at a series of points. Spot heights and surveyed contour lines thus give precise values of the height of the land surface and form quantitative data on the map. Contours are assumed to be accurate but errors can occur. If an error (dh) is made in height then an error is made in position (dl) which is related to the ground slope

$$\left(\text{since } \tan \alpha = \frac{dh}{dl} \right)$$

(Figure 2.2). Imhof (1965) has demonstrated a variety of possible contour errors arising from such instances of systematic errors in position or height, faulty contour intervals and accidental and gross errors. The majority of contours on present-day maps are compiled from stereoscopic air photographs and refinements in photography and photogrammetric techniques have led to a high degree of accuracy in their compilation.

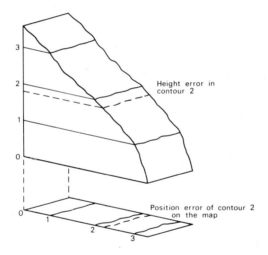

Figure 2.2 Diagram to illustrate the effect on the plan position of a contour line following errors made in height determination (or *vice versa*)

Contour lines are normally shown on maps and plans at regular intervals, the interval between successive contours being known as the vertical interval (VI) and being indicated on maps produced in non-English languages as 'equidistance' (or its equivalent). The appropriate spacing varies according to the terrain and needs of the map maker and map-user. In low-lying areas a frequent contour interval is clearly necessary whereas in hilly areas this could lead to confusion. Thus on a 1/25,000 map, if one allows for the printed line to be 0·1 mm wide and a spacing of 0·4 mm to be capable

Figure 2.3   Enlarged cross-section to show relationship between printed colour lines (base) and their true vertical interval, for slope of 45°

of being resolved by the eye between the centres of the lines representing two adjacent contours, then a contour interval of 12·5 m might be used even for slopes of up to 45° (Figure 2.3). An approximate formula that can be used for obtaining a possible contour interval is given below, assuming that a more realistic resolution distance is 1 mm.

$$CI = \frac{S \tan \theta}{2 \times 1,000} m$$

where S is the denominator of the map Representative Fraction and θ the steepest slope to be represented. For instance, with a 1/50,000 scale map:

$$CI = \frac{50,000}{2,000} \tan 45°$$
$$= 25 \text{ m}$$

Imhof (1957) has further shown that the ideal contour interval is in fact given by the formula $CI = n \log n \tan \theta$ metres, where

$$n = \sqrt{\frac{S}{100} + 1}$$

and S is the denominator of the map-scale Representative Fraction.

The contour lines shown for a given area often do not adequately depict all the relief features; for example, locally important hollows or terrace

Table 2.1 Contour spacing in metres for various map scales and for three different types of terrain (mountainous regions, slopes up to 45°; areas of medium relief, slopes up to about 26°; flat areas, slopes not above about 10°)

	Relief						
	Mountainous					Medium	Flat
Scale	a	b	c	d	e		
1/2,000	1	2	3	2	1	1	0·5
1/5,000	2·5	5	6	5	2·5	2	1
1/10,000	5	10	10	10	5	5	2
1/25,000	12·5	10(20)	19	20(30)	10(15)	10	2·5
1/50,000	25	20(25)	30	20	10	10	5
1/100,000	50	50	47	50	25	25	10(5)
1/250,000	125	100	85	100	50	50	10
1/1,000,000	500	200	200	200	100	100	20

a Smallest contour interval for ease of drawing (see Figure 2.3)
b Most commonly used contour interval (including alternatives)
c Theoretical contour interval (from formula on page 20)
d Recommended contour interval for principal contours
e Recommended interval for intermediate contours if needed
Alternatives are indicated in brackets.

features may not be represented at all because the selective contour interval omits them. This can be remedied by introducing additional 'in-between' contours at half or even one-quarter, half and three-quarters of the normal contour intervals. Such extra lines must, however, be in a different thickness or style of line and should only be used where necessary. It is also normal practice to label contour lines with their height value in such a way that the numbers stand with their base on the downhill side. Again, contour values may be placed on the line or, preferably, a gap made in the line. Thickening of certain contour lines helps in the interpretation of values but must be at regular interval (Figure 2.4).

RELIABILITY OF CONTOURS

The reliability by which contours have been mapped has increased with improved methods of surveying and of photogrammetry since the nineteenth century, with the result that the mean error of contouring on 1/25,000 maps has improved from $\pm(4 + 25 \tan \theta)$ before 1890 to

$\pm (0·5 + 5 \tan \theta)$ at the present time (that is, on slopes of $45°$ an accuracy then of ± 29 m, now of $\pm 5·5$ m, and on slopes of $10°$ an accuracy then of ± 9 metres, now of $\pm 1·5$ m). Table 2.2 indicates the accuracy of contours on some common topographic map series.

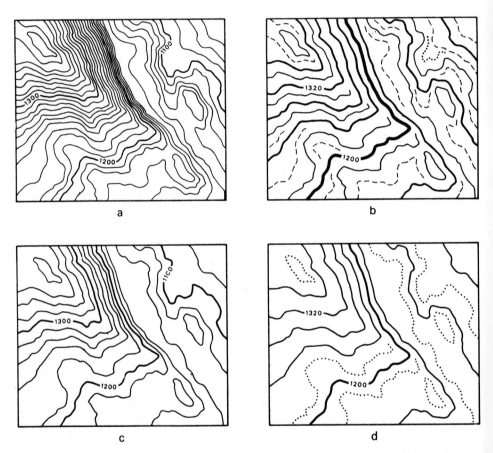

Figure 2.4 Examples of different contour intervals depicting the same relief pattern (after Imhof): (a) shows signs of coalescence of contours and hence at the scale of reproduction (c) would be most suitable. At a smaller scale (b) or (d) would be more satisfactory, (d) showing the better generalization. Note that the 40 m interval means that only certain 'hundred metre' contours will be present, i.e. in this case 1,200 m only.

   a Contours at 10 m interval, thickened at 100 m
   b Contours at 40 m interval, thickened at 200 m, and with intermediate local
      contours at 10 m and 20 m spacings
   c Contours at 20 m interval, thickened at 100 m
   d Contours at 40 m interval, with intermediate local contours at 20 m spacing
      only

Table 2.2 Table of formulae for calculation of errors in height of contours at various scales for some European map series (after Imhof)

Country	Scale	Mean error in metres
Germany	1/5,000	$\pm(0\cdot4 + 5\tan a)$
Switzerland	1/10,000	$\pm(1 + 3\tan a)$
Great Britain	1/10,560	$\pm\sqrt{1\cdot8^2 + 3.0^2\tan^2 a}$
France	1/20,000	$\pm(0\cdot4 + 3\tan a)$
Germany	1/25,000	$\pm(0\cdot5 + 5\tan a)$
Great Britain	1/25,000	$\pm\sqrt{1\cdot8^2 + 7\cdot8^2\tan^2 a)}$
Switzerland	1/25,000	$\pm(1 + 7\tan a)$
Switzerland	1/50,000	$\pm(1\cdot5 + 10\tan a)$
USA	1/50,000	$\pm(1\cdot8 + 15\tan a)$

QUALITATIVE REPRESENTATION OF HEIGHT

The pictorial representation of the relief of the land has been attempted by cartographers from the very early days of mapping. In most primitive maps a general side view of a 'sugar loaf' type of hill was given, at first as separate symbols, sometimes arranged along lines and occasionally in an overall pattern resembling fish scales. An attempt at the depiction of the actual shape of the individual hills was made by Leonardo da Vinci on a map of Tuscany dated 1503 but one of the first efforts at producing a three-dimensional impression of the landscape is that of Gygers' map of the Swiss Canton of Zürich made in 1667 at a scale of 1/32,000.

The drawing of lines *down* the slopes of hills so that the lines are closer together or thicker at the top and more widely spaced or thinner lower down had been practised for a long time but the formal introduction of *hachures* following specific rules dates from the nineteenth century.

The following rules should be observed in drawing hachures:

a They must be drawn in the direction of maximum slope.
b They must be arranged in rows, not drawn down the entire slope.
c The length of each stroke should indicate an equal vertical drop in height. Thus a series of short strokes will be drawn over a steep slope, longer strokes will indicate a gentler gradient.
d The thickness of each stroke must be constant unless a form of 'illuminated relief' is being presented.

A modification of hachuring, known as the Lehmann system (Lyons 1914) keeps the hachure lines at an equal spacing but the angle of slope is shown by the thickness of the lines. A definite relationship exists between the two which on the Lehmann system gives a completely dark cover on slopes of 45° or more. Other systems allow for a black/white gradation until slopes of 60° are reached. Yet again a system of, say, four 'tones' allows for some flexibility. The number of strokes per centimetre also determines the apparent density of hachuring (Figure 2.5).

Hachures may be applied indiscriminately to all slopes irrespective of their aspect but a much better impression of the landscape can be given if shading is introduced. Here the precise relationship to slope angle is lost, but if, for instance, slopes facing an apparent source of illumination in the north–west are shown with a lighter network of hachures than those to the south–east, an illusion of three-dimensional relief is produced. This form of hachuring can be termed 'shaded hachuring' (Figure 2.6).

Figure 2.5   Hachures: the Lehmann system in which width of line corresponds to slope angle for two different numbers of lines to the inch. The formula for width of hachure is given by the relationship

$$\frac{\text{width}}{i} = \frac{x^0}{450 - x^0}$$

where x is the angle of slope and i is the interval between the lines. Thus slopes of 45° will be represented by solid colour.

  a Slopes of 30°
  b Slopes of 20°
  c Slopes of 10°
  d Slopes of 5°

Many topographic maps in the nineteenth century used the system of hachures to depict relief and although the Ordnance Survey 1 inch to the mile maps may well be the best known to British readers there are only a few areas where extremes of relief make for difficulties in interpretation. By contrast, maps of some European regions of this time provided a challenge to the map maker, and those of the mid-nineteenth century of Switzerland find a special place in the history of cartography. They are generally referred to as 'Dufour' maps after the name of the Director, G. H. Dufour, and an example of one such map is shown at the top of the frontispiece to this book.

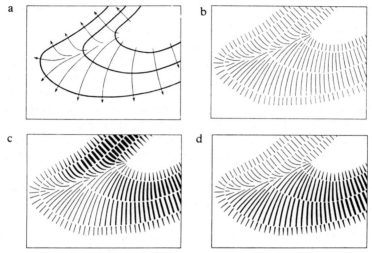

Figure 2.6   The principles of hachuring: the four diagrams indicate the manner in which hachures relate to the physical typography (based on Imhof 1965: 241, Figures 139–42).

  a The relief as delimited by contour lines and by slope lines: the latter are normal to the contours and are indicated by arrows pointing down the steepest slope.

  b The resulting basic arrangement of hachures to represent the same relief as in the previous diagram.

  c Hachuring in which the thickness of the hachures is related to the slope angle: thus the steep slopes at the top are shown by thick hachures.

  d Shaded hachuring in which a light is assumed from the top left (north-west) corner: hence southern and eastward facing slopes are represented by thickened hachures.

The method of using apparent illumination from the top left (north–west) corner is used in other methods of relief depiction. Apparent lighting from vertically above, so that valley bottoms are darkened and ridges left light in colour, has also been used but lighting from other presumed directions gives the impression of 'inverted relief'. An actual three-dimensional model of the landscape with an illumination from the top left will possess areas of light and shade that can be expressed in gradations of tone ranging from white to the full colour. If a colour plate is made either from a photograph of such a model or by the application of tones assuming the area to be lighted from one specific direction it is seen that the resulting tones grade smoothly into each other. Such tones occur in nature and are recorded in photographs but require presentation in printing processes by the use of 'half-tone' methods (see pp. 133–5). The resulting shading on the map is referred to as illuminated relief, plastic shading or hill shading, and can obviously be shown in any desired colour. Examples of its use are in maps at scales varying from topographical sheets at 1/25,000 and 1/50,000 (e.g. French, German, Swiss series), to 1/63,360 (Tourist sheets of the British 1 inch to the mile map), to small scale atlas maps produced in many countries.

Colour has been used in the depiction of relief for a number of years in the method known as 'layer tinting'. Such colour or shading is referred to as a *chorisogram* and is applied to the spaces on the map between the isograms, which in this case are the contour lines. Common usage in atlases and wall maps is of a system of colours grading from light to dark as the land increases in height. It has become established practice to have a range of colours through the tints of green, brown and red, with purples and/or whites on the highest areas.

Few maps use only one system for the representation of relief as described above, combinations of both quantitative and qualitative methods being generally adopted. Even on the largest scale plans where precise values are given in the form of spot heights, some additional symbols will be used for abrupt breaks of slope such as embankments or quarry faces. Here the hachure symbol or a modification of hachuring as in the cliff symbol, with sketches of rock outcrops, are used (Figure 2.7). Where a hill-shading technique is used it is usually necessary to print this shading in a separate colour, but it is possible to combine this with layer shading.

A particularly interesting combination of layer tints and hill shading has been used in a map of the Freetown area at 1/50,000 produced by the Directorate of Overseas Survey for the Director of Surveys and Lands, Sierra Leone (Carmichael 1969).

Figure 2.7 Symbols to represent steep irregular slopes where contouring is not possible: modification of the hachuring principles of lines down the steepest slope (b), or of horizontal hachures or form lines (a), can be used separately or combined as in (c).

DRAINAGE LINES

Lines of drainage are conventionally represented in the coloured topo-graphic map by blue lines, bodies of water by a blue shade, and informa-tion relating to water by lettering in blue. Other distinctions that are made relate to perennial and intermittent drainage (the latter usually shown with a broken line) and the change between riverine water and tidal water (that is, the highest point to which tides flow) often needs to be distinguished. Sources of water, springs, wells and so forth may or may not be shown, often depending on the map scale.

Blue is also the colour normally used by the cartographer for depicting details relating to ice, for example, limits of glaciers, contour lines across glaciers, icefalls and crevasse features of the more permanent type.

This convention is, however, not always followed when the features of the seas and oceans are to be shown. As was stated earlier the datum line for all height information is usually taken as mean sea-level. More significant lines on coastal margins are those representing high and low tide marks and, if the map scale permits, these can both be indicated.

Information relating to the shapes of the floors of lakes and oceans can be presented in much the same way as relief information but in most cases only submarine contour lines are used, layer shading being found in some atlases and 'hill' shading in relatively few specialized maps. The navigator's or yachtsmen's needs of maps of sea areas are somewhat different from the land-user and the mapping agencies serving them produce a specialized form of map: the hydrographic chart.

Limited hydrographic information is portrayed on most maps produced by topographic mapping agencies but is given in greater detail on specialist charts prepared by national hydrographic bureaux. Such charts fall into three main categories: ocean charts, harbour plans and navigational charts. The information shown on charts and plans will include details of water depths (in fathoms, feet or metres), nature of the sea bottom, position of reefs and hazards to navigation as well as data relating to tides and sea currents. Navigational charts are prepared on various map projections and generally have a close graticule network to assist with the plotting of position and courses. They can also be of an 'outline' variety covering a large area or may be designed for use with some navigational aid (for example, Decca systems).

Hydrographic charts show depth of water by printing soundings as spot values and also by inserting fathom lines – isobaths – or depth contours. These are related to a datum stated on the Tidal Information Table on the chart, the actual soundings normally having been 'reduced' to the level of lowest low water. Where a sounding is given its precise position is that of the middle point of the figure(s). On older charts a different symbol was used for each fathom line and various print styles for soundings of different degrees of accuracy. A full explanation of the 'Symbols and abbreviations used on Admiralty charts' is published by the Hydrographic Department.

Non-coastal information will be given only where relevant for navigational purposes, such as, for example, prominent landmarks. Land relief is shown by hachuring, broken contours or form lines. In order that all navigational information is up to date a system of continuous revision is available through the daily publication of 'notices to mariners'. Chart-users are required to make any necessary amendments to their charts from the information in these 'notices'. Where considerable changes occur the chart is withdrawn and a new issue produced. Publication runs of hydrographic charts are therefore small and stocks kept at low levels.

Until recently, charts were produced in a nearly monochrome style, the only colour being small patches of red or purple used for lighthouses and similar navigational aids. International specifications now include colouring, for example, yellow for land areas, and layer colouring to indicate the shallower water over continental shelf regions. This may be in one or more shades of blue and blue is also used to indicate marine contours (isobaths). The metre is now the international unit for depths and the fathom unit is disappearing from use (Figure 2.9).

Most British charts were formerly prepared from engraved copper plates and, in recent years, they have been photographically transferred to zinc-enamel for offset lithographic printing. The 'engraved' style is therefore still current in such charts and is particularly noticeable in the profiles added to give an indication of the mariner's view of the coastline (Figure 2.8).

Figure 2.8   Illustrative sketch formerly placed on Hydrographic Charts to assist in the recognition of harbour entrance courses (extract from Isles of Scilly chart)

Figure 2.9 Extract from Hydrographic Chart: the original has colour shadings on land areas and areas exposed at low water and to show other features. The scale is approximately 1/15,000 in this reduction. Depths are in metres.

VEGETATION

The mapping of vegetation distributions necessitates areal symbols of some form (McGrath 1965 and 1966). The requirement can be met by means of a repeated single symbol, for example that representing a tree or bush which on the very largest scale of map will be a precise locational representation of, for instance, individual trees. As the map scale reduces, the exact positioning of each object will not be possible because of the size of symbol used, since the symbol must remain large enough to be recognized easily (Figure 2.10). A generalized pattern of symbols will then be used or, alternatively, shading or colours to represent the distribution of the vegetation under study. These are limited on most topographic series to a few

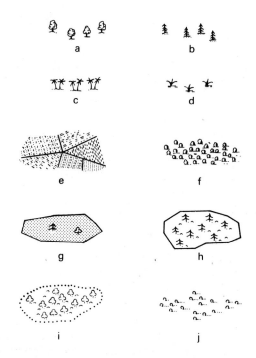

Figure 2.10 Some vegetation symbols used on maps and charts (a–f from Admiralty charts, g–j from Ordnance maps)

a Deciduous trees
b Conifers
c Palm trees
d Nipa palms
e Cultivated fields

f Bush growth
g Parkland with some tree growth
h Fenced coniferous woodland
i Unfenced deciduous woodland
j Scrub undergrowth

broad categories of natural or semi-natural vegetation such as heathland, rough pasture, woodland. Shading, symbols and so forth may also be used to represent distinctive cultivated land-use categories such as plantations or orchards. One feature of the British landscape that has given rise to a symbol that may be considered to belong to the 'vegetation' category is that used for parklands, both public and private, which are shown by stipple on Ordnance Survey maps of 1/250,000 and larger scales.

3
Representation of the activities of man

The mapping of all topographic details of the earth's surface leads to problems of representation and of generalization. These are a result of the different types of man's *occupance*, for example agricultural land use, settlement and communications, and they also stem from different *scales* of map representation. Cartographic problems arise, too, in representing information both on single-purpose thematic maps and on multi-purpose topographic maps. One must consider matters such as statistical representation, colour choice, balance of tones, use of symbols and lettering.

In all cases, however, the information depicted falls into one of the following categories:

a LINEAR information, e.g. boundaries or communication lines, representing either dividing lines or flows and movement.

b AREAL information, which usually leads to map shading or colour tones, e.g. individual buildings at large scales, entire settlements at small scales, land use information at all scales. This is also referred to as 'patch' data.

c SYMBOLIZED information, which can be information relating to a POINT where a conventional shape represents a feature in a pictorial or other manner, e.g. geometrical shape, or in the form of a statistical diagram, bar-graph, divided circle, etc. relating either to a point or to an area.

d WRITTEN information in words, abbreviations, numbers, etc. This can relate to a point, area or line.

All four categories indicate a representation of the topography in some 'conventional' form and lead to the necessity for a map *legend*. This is sometimes referred to as the map 'key' but this term has other technical connotations in cartography (i.e. the base outline of the map before colour separations are prepared in map reproduction – see Chapter 12).

Map information can be presented in any of the above methods either generally, merely to give qualitative information as to location or activity (town name, type of industry), or more precisely. The latter will give quantitative information and factual data (for example town population or industrial production for a given date).

Another factor common to all the categories is that the information shown on the map may not physically be evident in the topography. Thus a pipeline may not yield a continuous line on the ground, and for this reason may not have been mapped by the land surveyor. Place-name information and administrative boundaries are further examples. Lettering or explanatory words may well be required to supplement or to explain other symbols on the map (for example evidence of former land use or occupance: prehistoric settlements and burial sites, forts and tumuli).

The cartographer will aim to depict as much of the ground information as possible at the scale to which he is working and will need to balance possibly conflicting needs of the map's purpose, map symbols, written information and so on. Over many years a number of particular symbols and colours have become accepted or 'conventional' usage but there have also been changing fashions in map representation stemming from advances both in the complexity of the topography and in cartographic techniques or processes of reproduction.

LINEAR INFORMATION

Printed lines on the surface of the map represent a whole series of topographic features. However, lines can generally be taken to represent boundaries in some form. Where such boundaries are clear and definite they will be shown by a full line, for instance fences, walls, edges of buildings. Where low and indefinite (road/pavement edges, unfenced fields) they will be given a broken line. Thus on the largest scale maps a road will be indicated by full lines for the hedges, together with broken lines for pavements. Two printed lines will be present also for roads at smaller scales, probably with a colour fill between them, but one or more of the lines may be broken to indicate lack of fencing, or road category. Eventually, however, a single line of colour may be used to represent the road or other communications line. Even without the use of colour it is possible to utilize a wide range of line symbols by introducing broken lines, chain lines or additional symbols along or across the line (Figure 3.1).

All these examples, whether for roads, railways, canals, pipelines, transmission lines, administrative boundaries or even for the distinction between hedges, wire fences and boundaries with trees, are of a qualitative or purely descriptive nature.

In order to introduce a quantitative element in a line symbol it will be necessary to provide it with a dimension. At its simplest this can be a written figure alongside the line but on thematic maps showing flows along route lines a variation in line-thickness is the usual practice (Figure 3.2). In cases where a large variation in values is to be catered for, it may be necessary to use a pictorial device whereby the line becomes an apparent three-dimensional shape, such as a cylindrical pipe or one of rectangular cross-section.

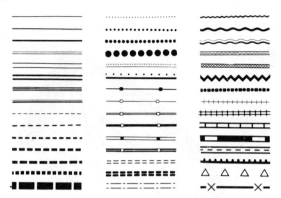

Figure 3.1 Line symbols: examples of different symbols that can be used for lines, e.g. boundaries, communication lines, flowlines. Most of these are available in pre-printed form, as tapes that can be applied to drawing surfaces (see p. 125); the possible range of linear symbols is far more extensive than can be shown here.

AREAL INFORMATION

Many activities in the field of human geography can be considered under this heading: settlement, industry, agriculture, actual land use, potential development and so on. The larger the scale of map at which the topography is to be represented the more easily can the information be presented by the scale plan form of the activity, by, for instance, individual houses, industrial sites and land parcels. Particular uses or activities can be shown by a shading or colour placed over the area concerned. Some examples of the various shadings used on maps are given in Figure 3.3. This shading can indicate the activity in a purely descriptive way as on land-use maps, or it can give numerical information according to a pre-determined shading or colour legend. Such shading/colours are referred to as *chorograms* and where the shading/colour is applied to a statistical or administrative division the term *choropleth* is used.

The term *chorisogram* has already been defined (see page 26) where shading is applied to the intervals between measured isograms (lines along which a value is constant). In thematic mapping other lines may be drawn merely to join points having the same value, and these are called *isopleths*. Shading between such lines is therefore technically referred to as *chorisopleth*. Similarly if the isolines were measured accurately they should be termed 'isometric lines' and the shading between them would be referred to as a *chorisometer*.

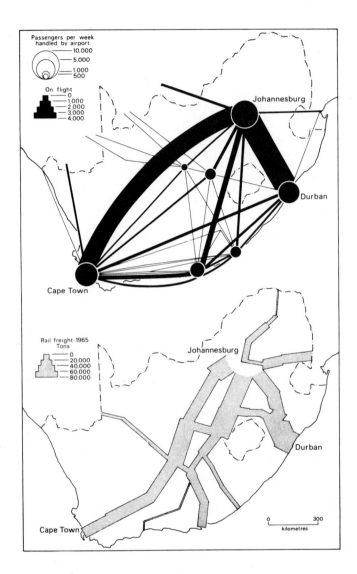

Figure 3.2 Flowline maps: the two map examples given here show respectively (top and bottom) domestic weekly passenger air traffic (1966) and main-line rail freight (1965) in South Africa. The width of the line is proportional to the value and thus requires a legend in addition to the circle scale for airport passenger handling in the top diagram. Less important connections are not dimensioned. Note that in the lower diagram the flowlines follow the actual routes and also vary along their course. (After Board *et al.* 1970: 379–80, Figures 11 and 12.)

Problems that occur in any such quantitative representation of data can be considered under the headings of the data themselves and the choice of shading. Problems provided by the data generally stem from the sources from which they come. In the published thematic map it can be assumed that the data are accurate and adequate and the only problem is that of interpretation. The greatest importance therefore lies in the map legend. Figure 3.4(a) and (b) shows two methods of representing the same data; in the first case no account is taken of the statistical distribution of the values while in the second the shading refers to one or more standard deviations from the mean. Alternatively, these figures could have been mapped by reference to the position of, for example, the mean, upper and lower quartiles. A discussion of these particular examples is to be found in the *Cartographic Journal* (Storrie and Jackson 1967) and there is a further discussion of population mapping together with a number of maps in *Transactions of the Institute of British Geographers* no. 43, 1968 (Hunt 1968).

SYMBOLIC REPRESENTATION

The simplest methods of representing distributional data are those of dots and proportional geometrical shapes. The dot is clearly the simplest symbol that can be used and, basically, one dot can represent *one occurrence* (single or group) of the data to be mapped. Usually the single dot will stand for

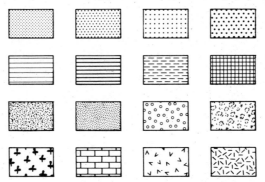

Figure 3.3 A small selection of the many monochrome shadings possible for cartographic use: all the examples shown are from pre-printed sheets and the top two rows show possible ranges that could be used against a numerical scale. The proportions of dark to light can be related precisely to given values if required, by varying line widths or dot sizes and spacings. The bottom two rows show a variety of area-shading symbols that could be used as *choropleths*.

a number of items and the dot itself may be replaced by some other simple shape either geometrical or pictorial. Ideally it should be possible to count the number of symbols and so obtain the value mapped. The distribution of these symbols can be made in a uniform manner – as for vegetation in-

Figure 3.5 Thematic map of population changes in England and Wales 1951–61 (based on Osborne 1970): this map indicates a method of combining proportional circles and choropleth shading. The circles are drawn according to the 'square root' principle in which areas are proportional to the quantity to be mapped. It is possible to scale off each circle and compare it with the extended scale shown at the base of the map, but the visual impression may not be as distinctive as that in which an exaggeration has been used – for example the so-called 'Flannery' method given by tables in Robinson *et al.* (1978) that is indicated in Figure 3.6.

The shading is on a regular interval basis and distinguishes between positive (line shadings) and negative (dotted shading) values. All the shadings are capable of application by hand or by pre-printed methods and in fact a combination of the two has been used here.

Figure 3.4 Representation of statistical data: these two maps present the same information, the proportions of owner-occupiers by counties in the British Isles, in two different ways. In (a) the data are those of actual percentages, with equal divisions of the 0–100 scale (no values below 12·50). Note that there is no overlap of value, i.e. the divisions are 12·50–24·99, 25·00–37·47 and so forth. (b) presents the same data in terms of deviations from the mean, the range here being across five standard deviations. Shading has been applied to both these maps such that there is a distinction between values below and above the mean value (44·52 per cent), and in (a) the category containing this figure is shaded distinctively.

formation, described on page 29. It is more usual, however, for the symbols to be placed in geographically correct positions; for example, dots for population distribution will be clustered in urban localities and scattered over remote hilly areas. Such plotting of a distribution is referred to as *dasymetric plotting*. Once the data to be represented exceed that which can satisfactorily be shown by a number of single symbols it is customary to use a geometrical figure that can be measured against a given scale. The most usual figure is a circle and in order that this may correctly represent the required data the *area* of the circle will be made proportional to the figure it is to show. The legend to such a thematic map will then include a scale as in Figure 3.5, against which the map-user can measure the symbol by means of a pair of dividers.

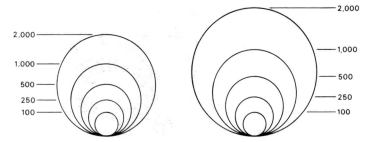

Figure 3.6 Proportional circles: when circles are drawn to represent data it is usually regarded as satisfactory to make the *area* of the circle proportional to the data (left-hand diagram). However, the impression of relative sizes is not entirely satisfactory and it is better if an enlargement is made (right-hand diagram). A table giving values for the radii of proportional circles is given in Appendix G of Robinson, Sale and Morrison (1978)

WRITTEN INFORMATION

By far the easiest way in which to record and present factual information on a map is by means of a printed statement in words and/or figures. The simplest and most obvious examples are place-names and values for ground height. Provided the map scale is large enough, very much more information can be recorded in this way. Indeed, the field surveyor of, for example, agricultural land use will probably proceed by annotating the large scale plan. For complex patterns, such as those in urban areas, it may be necessary to record only reference numbers on the map and list details of building, occupance and other data on a separate booking form.

Most names on published topographic maps are place-names. In order to distinguish between places of different size or of different status (for example in terms of administrative function, capital cities, towns, parishes and so on) a different size and style of lettering can be employed (Figure 3.7). Also, variations in lettering can be used to indicate different historical ages of the topographic features. In order to overcome the extravagant use of space on the map surface that lettering can involve, abbreviations are frequently introduced. These will usually refer to a given location, normally marked with a point and the full legend to the map will include a comprehensive list of all these abbreviations (see Figure 4.1).

Numbering on maps includes such information as spot heights and, on large scale plans, reference numbers to the individual land 'parcels'. On British Ordnance Survey maps at the scale of 1/2,500 these parcel numbers are quoted together with a statement of the area of the parcel (Figure 3.8). Various methods of presenting this information have been used and these are summarized in the OS publication (1968) *Parcel Numbers and Areas on 1/2,500 Scale Plans.*

Other numbers that can occur on topographic maps are those relating

a

Over 500,000	**LOS ANGELES**
100,000 to 500,000	**OMAHA**
25,000 to 100,000	**GALVESTON**
5,000 to 25,000	**Laramie**
1,000 to 5,000	**Grand Coulee**
Less than 1,000	Sun Valley

b

Parish Churches & Villages	Bridges *(On Main Roads)*, Bridges *(Other)*
Other Villages	*Isolated Houses*
PARKS & DEMESNES	BAYS & HARBOURS
Gentlemen's Seats	*NAVIGABLE RIVERS & CANALS*
Manufactories, Mines, Farms, Locks	*Small Rivers & Brooks*
Local Authority Establishments	BOGS, MOORS & FORESTS
	Woods & Copses

Figure 3.7 Examples of lettering styles and sizes used to indicate different features on maps
    a Quantitative use, where size of lettering indicates approximate size of detail, e.g. town size by population (extract from US 1/250,000 map legend)
    b Qualitative use, where different printing styles indicate different features (extract from Conventional Signs sheet for Ordnance Survey 1/2,500 (County sheet-lines)): current lettering styles on Ordnance Survey plans are less varied than this. Administrative areas on the 1/1,250 series are indicated in the margins and the lettering is designed to be capable of enlargement without excessive emphasis where the base map is the 1/1,250, since this forms the basis of the 1/2,500 map.

Figure 3.8 Parcel numbers and areas on Ordnance Survey 1/2,500 plans

1 Bracing symbol to indicate areas combined
2 Centre bracing to indicate that area is measured to the centre of this strip
3 Open bracing where parcel is crossed by a second feature: road over river and islet joined with field
4 Symbols to indicate the edge of the built-up area: such an area will include a large number of small parcels that would be difficult to show separately, therefore all inside the region defined by this symbol will be summed to give a total area
5 Reference number (grid reference to nearest 10 of approximate centre of parcel) and area of parcel: the area is given in hectares (abbreviation ha) and acres, the former to three decimal places and the latter to two.
6 Parcel number and area as far as the edge of the map
7 Corner parcel number and area where parcel extends beyond map corner
8 Parcels in separate parishes
9 The parcels here are sections of the road and river respectively.

Additionally, this extract gives some indication of the degree of detail on plans at this scale. Buildings are shown by their ground plan (although minor details may be generalized), house numbers are indicated and the edges of pavements are shown. The woodland detail, however, is symbolized. The word 'issues' and the arrow indicate a drainage line or field ditch.

to population (on the French 1/25,000 and 1/50,000 series the figures are given after the printed place-name) and such information as distances along roads between marked points, and the official road classification number. In the map margins distances to settlements on adjoining sheets are often given.

DISTRIBUTIONAL INFORMATION

Thematic maps can present geographical and distributional information in the variety of ways considered above but it is often necessary on such maps to show or examine statistical data for an area rather than for isolated points. In addition to the choropleth methods or the presentation of figures and symbols on the map the cartographer may use a method allied to contouring for the depiction of relief. This is the technique of drawing *isolines*, lines joining points of equal value. In contouring, the actual land surface can

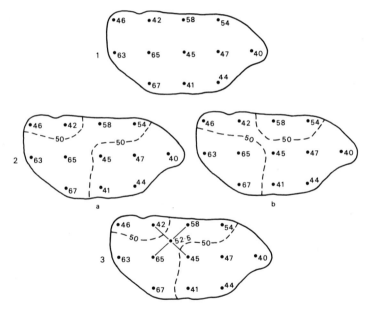

Figure 3.9 Plotting isolines from grid data (1): in most cases this presents no problem but in this particular example two solutions (2) seem possible. If the four surrounding values are averaged (3) a value is obtained for the central point and this provides a solution or alternatives. The position of the isolines in the remainder of the example is obtained by assuming a steady change in value between any two points (Board 1967, after Mackay 1953). There are cases, however, where it is necessary to choose between two possible solutions on the grounds of the information to be conveyed. Thus, if the appropriate map were to be used for air navigation, where ground heights are critical, a solution should be given indicating the highest possible value from the data (i.e. solution 2a is to be preferred). On the other hand, if the values relate to critical depths, as on a hydrographic chart, a solution that shows the minimum depth is preferable, i.e. 2b.

be measured and the precise position of the contour so obtained. In other distributions it is normally only possible to know precise values at a network of points. These points may be positioned randomly (random in either a literal or statistical sense) or placed regularly according to some network or grid. In either case it is likely to be necessary to interpolate the position of the isoline from a network of spot values. An assumption must be made as to the rate of change of the data between the given values and in most cases this will be taken as being at a uniform rate. A method of calculating the position of isolines for given spot values is shown in Figure 3.9. The drawing of isolines and also the plotting of choropleths can be undertaken by various computer programs. Many other ways of representing distributional data are now available as a result of computer mapping systems (Rosing and Wood 1971) or from data banks (Hägerstrand 1967).

4

Map types

The distinction between two major map types – topographic and thematic – has already been made. Published map series, however, fall into a number of distinctive types that reflect the basic mapping need and also the 'house style' of the map-producing agency. This style is a result of the particular specifications to which the map publisher has worked and is due to the combination of controls of layout, lettering, colouring and additional information, as well as the topographic and/or thematic information portrayed on the map itself. The method of reproduction provides a further factor in establishing the map style.

Distinctive map styles characterize not only the two main groups of maps so far defined but more particularly the organization producing the map. The official map series of different countries thus show variations ranging from the shape and size of the map sheet to the way in which some topographic features may be presented. Careful reference to the map 'legend' must therefore be made. The legend printed on the map itself may not be the full legend adequate for the entire map series (for example the British and Swiss 1/25,000 map series) and in such cases it will be necessary to obtain the separately published legend. In many European cases the legend is multilingual, for example the French, Dutch and English of the Belgian and Dutch 1/25,000 series (printed on the map); the French, German and Italian of the Swiss map series; and the Russian and English of some Russian maps, particularly the more readily available small scale maps.

OFFICIAL MAP SERIES

The official topographic map series of any country, although with differing styles, usually form a distinctive group in which both monochrome ('outline') or fully coloured editions are available depicting relief, settlement, communication and some vegetational information as well as names and political boundaries. The larger scale maps prepared for cadastral purposes are usually in a single colour, black or grey line work being the most common. Planning maps, on the other hand, are more usually produced with colour shading. Both these categories are the work of official mapping bodies and bear the distinctive style of that body. As scales increase, the monochrome cadastral maps become more and more alike in appearance, but the thematic maps of planning bodies often reflect the stamp of the printers and publishers.

Figure 4.1 Extract from Dutch map legend: a small portion of the legend to the Dutch official topographic map series at 1/50,000 scale to show how a multilingual legend can be set out

COMMERCIAL MAPS

Maps produced by non-official agencies usually rely heavily on the various official topographical map series. The maps that such bodies subsequently produce are often of the thematic type and fall into such categories as:

1 Town plans at various scales
2 Road maps at various scales
3 Itineraries – or maps of particular routes – and location or site maps
4 Tourist maps of specific regions
5 Miscellaneous thematic maps, e.g. marketing, distributions and productions
6 Network maps of rail, air or other communication patterns

Nevertheless, commercial bodies also produce topographic maps, sometimes as individual sheets and sometimes as a series giving national cover.

SEA AND AIR CHARTS

Maps that fall somewhere between the specialized thematic maps listed above and the official topographic map series are those relating to sea and air communications. Most maritime countries produce their own hydrographic charts (see page 27) not only of adjacent coastal areas but also covering overseas waters, so that sailors may have a chart printed in their native language. Many of the latter are derived from the work of other hydrographic surveys.

For aeronautical purposes a further range of special maps or charts are required, and here the special problem is that of keeping up to date, as restricted areas, flight lanes and so forth may be subject to changes at short notice. Overprinting of symbols on a standard topographical map is a possible solution to this but the normal method is the production of a special 'air' edition making use of standardized symbols and shadings as laid down by the International Civil Aeronautical Organization (ICAO).

Both hydrographic and aeronautical map-users also require specialized charts relating to radio and other navigational aids. Here the usual practice is to utilize a base map with essential outlines on to which is printed the appropriate grid lines of the radio system, as, for instance, on the Decca, Loran and Consol charts.

In the succeeding chapters the official map series and some of the more widely available commercial maps produced in both the United Kingdom and some Western European countries will be briefly considered.

AERIAL PHOTOGRAPHS AS MAPS

Aerial photography has been used for several decades in the preparation of topographic and other maps, but usually as a survey method by means of the principles of photogrammetry. The air photographs themselves can, however, be substituted for the map in a number of ways with varying degrees of precision. For rapid mapping purposes a set of photographs may approximately be placed in positions controlled by a skeletal survey framework. This arrangement of photographs can itself be photocopied and produced as a 'print laydown'.

Alternatively the photographs can more carefully be joined and again, controlled by a skeletal survey framework, copied in a *mosaic* form. These mosaics may exist in both an unrectified and a rectified form. The latter indicates that correction has been made not only for the various distortions due to aircraft tilt and so forth, but also for scale.

No allowance can be made, however, for differences due to height variations on the photographs and, unless the area considered is perfectly level, the scale will vary with height. This can be overcome by the methods of *orthophotography* producing an *orthophotomap* (Scott 1969 and Zuylen 1969).

The photograph, consisting of various tones, can also be broken down into various component parts of light and shade of which the darkest tones are the shadows that outline areas. A series of masks and screens can be used to differentiate these parts of a photograph and the resulting process is known as PICTOMAP (Photographic Image Conversion by Tonal Masking Procedures – US Army Map Service).

These processes lead to the production of maps in which the air photograph base is clear to a greater or lesser extent and which can be termed *photomaps*. These are produced at a variety of scales, examples ranging from 1/5,000 in Nordrhein-Westfalen to 1/50,000 in Saudi Arabia. In most cases some interpretation is made by adding linework to the base photographs – to show contours, and to emphasize roads, houses, boundaries, and so on. Names, too, must be added and separate tones or land-use interpretations can be put on by the use of colour, as with the Swedish Economic Map at a scale of 1/10,000.

COMPUTER MAPS

Maps prepared by computer are increasingly becoming available and there are a number of computer programs that may be used to present geographical and other data in diagram form. Some of these programs have been prepared for particular purposes; others can be used with a variety of data or for various methods of analysing data and presenting the results in map form.

Maps produced by computers are generally prepared directly on the lineprinter output sheet. This is a continuous roll of paper that is folded into pages approximately 390 mm × 280 mm. The computer prints one line or row of characters at a time – hence the term lineprinter – and the characters are those of the conventional typewriter keyboard. Overprinting of characters allows for the printing of symbols of different intensity (Robertson 1967 and Rosing 1969). Each character occupies an area of $\frac{1}{10}$ inch on the X axis and either $\frac{1}{6}$ inch or $\frac{1}{8}$ inch on the Y axis (Figure 4.3). Each symbol is printed within this rectangle and a narrow margin of blank paper will result between each symbol giving a network of white lines over the map. Note that this shape is *not* rectangular as is the normal map grid. Linear maps can, however, be produced from computer data in co-ordinate form (Figure 10.2) (Diello *et al.* 1969).

Amongst the programs available for the preparation of maps by computer are:

1 SYMAP (Synagraphic Mapping System, Harvard University)
2 LINMAP (Ministry of Housing and Local Government, London)
3 COLMAP, a colour map system based on LINMAP
4 MAPIT, a program for the production of flow maps, dot maps and graduated symbols.

Figure 4.2 Orthophotomap example and comparison with topographical map of same area: the Orthophotomap is taken from an experimental map of Farringdon area (Berkshire, UK) produced by Fairey Surveys Ltd. The section of the Ordnance Survey map of the area has been reduced from 1/25,000 scale: grid lines at 1 km spacing.

Figure 4.3 Detail of print-out from SYMAP program, showing arrangement of squares in lines (eight per inch vertically) and columns (ten per inch horizontally): shading variations are produced by overprinting of type characteristics but, because the ordinary type style is used, small white lines surround each rectangle.

Figure 4.4 Grid pattern used by GRIDMAP computer programs: the main squares conform to those of the National Grid, but these are subdivided into rectangles to conform to the computer output (after Gaits 1969).

Based upon the O.S. Map

Figure 4.5 Computer map published in a regional atlas: shading obtained by over-printing (see Figure 4.3) to give a range of eight equal steps to 100 per cent; roads prepared and added separately with outline, legend numbers and scale. (From Rosing and Wood (1971), Map 14, 'The percentage of dwellings rented from local authorities'.)

SYMAP can be used with data relating to given co-ordinates to prepare either an isopleth or choropleth map as required. LINMAP was developed in the UK in order to operate a 'data bank' of given statistics relating to the General Register Office Census information at ward and parish levels. The maps so produced can be dotmaps, grid maps (shading related to a specified grid pattern on the map), zone maps (a form of choropleth map) or surfmaps (from surface mapping), in which values are calculated by interpolation from the sample values provided (Gaits 1969). The relationships between the National Grid system and the computer map are clearly exemplified by the GRIDMAP output, which is in a 5 × 3 or 10 × 6 format and which must be made to conform to a square grid as shown in Figure 4.4.

The basic data for computer mapping is related to a rectilinear grid such as the Ordnance Survey national grid. This data is then stored on tape and can be used to prepare a key map on a grid *square* basis (unlike the normal computer output described above). In the case of COLMAP the key map is replaced by a series of separate colour plates for litho printing in a conventional manner (*Penrose Annual* 1971).

5

British map series

There are many different map types produced in Great Britain for both general and special purposes as well as probably an even larger number of individual maps of specific areas. These may conveniently be grouped into two categories, both of which include examples of topographic and thematic map types:

  a Maps produced by official mapping agencies and government bodies, such as the Ordnance Survey
  b Commercial mapping firms: these produce maps of many overseas areas as well as of Britain.

In general, maps produced by the second group in Britain rely on base data from official surveys but for overseas work may undertake their own surveys.

THE ORDNANCE SURVEY

The Ordnance Survey was established at the end of the eighteenth century by the detailing of military personnel under Major-General Roy to undertake a 'Trigonometrical Survey' of Great Britain and Ireland. For defence purposes this work was under the charge of the Board of Ordnance – hence the name of the survey. It was eventually transferred to the Office of Works and in 1890 to the Board of Agriculture, but since 1967 has been the responsibility of the Ministry of Housing and Local Government (since 1970 under the Department of the Environment).

The original offices of the Ordnance Survey were established in the Tower of London in 1791 and the first published topographic sheet – the 1 inch to the mile map of the county of Kent – appeared in 1801. Larger scale maps were produced later in the nineteenth century, starting with a series of 6 inch to the mile maps of Ireland and later of Great Britain. The first maps at the scale of 25 inches to the mile appeared in 1853. The early large scale mapping of Great Britain continued at these two scales, the larger one being used only for the major built-up areas. However, these maps were on a county by county basis both as regards their projection and sheet-lines. Problems of discontinuity between counties or blocks of counties were considerable and the Davidson Committee on the Ordnance Survey reporting in 1938 recommended a number of changes. These were not fully implemented until the period after the Second World War and the final pattern of mapping is described below. Changes due to the introduction of metric units during the 1970s and 1980s are also noted.

Ordnance Survey map series

The largest scale topographic maps of Britain usually available are at the scale of approximately 50 inches to the mile, the representative fraction being 1/1,250. These maps are produced for towns and urban areas with a population of about 20,000 or more and are square in format, each sheet covering an area of a quarter of a square kilometre. The sheet-lines are based on the squares of the National Grid.

The old county 25 inches to the mile maps have been superseded by maps at a scale of 1/2,500. These will eventually cover the whole of Great Britain apart from areas of mountain and moorland. Some of these map sheets have been produced in a rectangular format covering an area on the ground of 2 kms from west to east by 1 km north to south and this will be the eventual pattern for all maps in this series. The information shown on both these maps is purely planimetric together with spot heights and heights of bench marks. In metric units the former are to one place of decimals and the latter to two decimal places. Areas are quoted at present to two decimal places of an acre but with metrication this has been changed to three decimal places of a hectare. As an interim measure values in both acres and hectares are shown.

Maps at a scale of 1/10,560 (6 inches to the mile) have been published for the whole of Great Britain. National Grid sheet-lines are used for the whole of England and Wales as well as most of Scotland but for remote highland areas it may still be necessary to consult the original county sheets which may not be contoured. Nevertheless, maps at the 6 inches to the mile scale are the largest for which a considerable amount of relief information can be obtained, the contour interval varying from 25 ft on the fully revised sheets to 50 ft on sheets of a 'Provisional Edition' that was compiled by redrawing from the original county maps together with additional revisions. After metrication this series will appear at the scale of 1/10,000 with contours at 10 m intervals in the more mountainous areas, 5 m elsewhere. It is likely to be many years before the entire country is covered at this new scale and a number of different map series will therefore co-exist for some time. The Ordnance Survey policy of 'continuous revision', however, will ensure that any major changes in the landscape will be recorded and reasonably up-to-date maps will remain available. All the above maps are based on the largest scale field surveys of the appropriate areas and have much in common. Generalization is needed only to a very limited extent and is confined mainly to roads and building outlines at the 1/10,560 scale. The result of such 'widening' of roads, however, is to make inaccurate any measurements of area by geometrical methods. The contours on this map are produced photogrammetrically and detailed height information is confined to levelled spot heights, quoted to the nearest foot or one decimal of a metre. For greater information relating to heights at this scale it is necessary to turn to the printed lists of bench marks that are prepared for every 1 km square of the National Grid.

At smaller scales only one series at present is confined to exact large National Grid squares. This is the 1/25,000 series, popularly called the '$2\frac{1}{2}$ inches to the mile' map, each sheet of which covers a ground area 10 kms by 10 kms. (The second full edition of this series, however, will be in a rectangular format of 20 kms west–east by 10 kms north–south.) Colour printing makes its first full appearance at this scale as apart from brown or red contours on various 6 inch to the mile maps, monochrome black is the dominant colour of the larger scales. Four colour printing is used on the 1/25,000 maps of the first edition which show most land parcels, contours at 25 ft intervals (to be 5 km and 10 km in lowland and upland areas respectively with metrication), generalized building shapes and woodland. Brown is used for the contours and road filling, blue for water, green for vegetation information and black for buildings, field boundaries and so forth. A fine black hatching provides a tone over buildings and the grid lines are also in black.

The 1 inch map

As well as being the first map type produced by the Ordnance Survey this was until recently the most well known of the British national maps and its history spans two centuries. For most of the time the maps appeared in a number of successive *editions*, each with a characteristic style. Thus the First edition was a monochrome black in which relief was depicted by hachures. This was followed, in 1840, by maps of the 'New Series', based on the 6 inch to the mile survey, and ultimately in 1893 by the Third edition. Different arrangements of the map sheets (or *sheet-lines*) were used for these editions. The layout of the Third edition remains today (in its small sheet-lines version) in the sheets of the 1 inch to the mile map of the Geological Survey (New Series) where it has been used in a black outline form as a base map.

The Fourth edition began in 1919 on different sheet-lines and like the Third edition was a fully coloured map. The Fifth edition, which was started in 1928, marked a considerable change in style, particularly in lettering and also in the introduction of additional methods of relief representation – hill shading being the most notable. This map series was never completed and when the need arose for maps during the Second World War the Fourth edition was revived in the form of the 'War Revision' series. From 1946 onwards the sheet-lines of the 1 inch to the mile map became more precisely related to maps at other scales by making these lines follow grid lines of the National Grid. This map was referred to as the Sixth edition and also as the 'New Popular edition with National Grid'. It was confined to England and Wales, Scottish maps being separately covered by the Scottish Popular edition based on the older Fourth edition.

The Seventh *series* 1 inch to the mile map first appeared in 1959. This was basically on the same sheet-lines as the Sixth edition but included the whole of Great Britain. However, there were a number of minor changes in style since its original publication. Some of these were concerned with the number of colours used, which have fluctuated from ten down through eight to six. Different editions within the series occur for each sheet, depending on the state of revision. For example, new roads and motorways may be added, or information relating to public rights of way over footpaths and bridleways may be made available. This involves some modifications to the map which are recorded in two ways. Firstly, a reference to the change is incorporated in the footnote outside the map margin such as: 'Reprinted with the addition of major new roads, 1970'. Secondly, an edition letter is also given to each revision of the map and this is shown in the bottom left-hand margin by 'A', 'B', etc. Such edition letters and footnotes are not confined to the 1 inch to the mile map; the appropriate edition letter is printed both in the top right-hand and the bottom left-hand margin of maps at larger scales.

The 1/50,000 series and metrication

With the national move into metric units in the 1960s one of the main arguments for the retention of the '1 inch to the mile scale' on grounds of familiarity and standard units in common use was removed. The full process of 'metrication' involves not only the adoption of metric scales but also the incorporation of metric units throughout the published range of maps, the substitution of metres for heights and a suitable contour interval, in metres.

The best replacement for the Seventh series 1/63,360 map, which to many users typified 'the Ordnance map', was ultimately agreed as the 1/50,000. This scale is used widely in other countries and its literal scale of 'two centimetres to the kilometre' is simple. Although the introduction of a new map series makes necessary the consideration of a whole range of detailed specifications, including sheet-lines and map layout which can vastly change the appearance of a map, the 1/50,000 map shows its parentage in the 1 inch map quite clearly. As an interim measure many of the Seventh series characteristics were merely transferred by a mixture of photographic techniques and some relettering to produce a First series on new sheet-lines. This was produced in two blocks in the 1970s (a northern and a southern one) and the interim First series will gradually give way to the definitive Second series over a period of some twenty years.

The 1/50,000 map is basically a six-colour map with a greater use of blues, greens and yellows than hitherto. An unusual feature, compared with preceding maps and maps of other countries, is the use of the blue 'fill' on motorways. Until it is possible to redraw contours at metric intervals the existing 50 ft lines are used with equivalent metric values. Some of these can be confused if read upside down and they are only used on broadly south-facing slopes with the consequent need for additional labelling of other contours. Various sheets of the 1/50,000 have also been selected to carry tourist information, which is additional data overprinted in blue, and some of the legend is given in three languages – English, French and German. Information on public rights of way that had gradually been added to the Seventh series 1/63,360 map is also shown but vegetation has been simplified by using a green tint without overprinting deciduous/coniferous black symbols, although there are different intensities of green between these two categories. Perhaps one of the most fundamental changes introduced by the Ordnance Survey in the 1/50,000 Second series – and elsewhere on maps at other scales – is the change to a more modern typeface, that of Univers. Nevertheless it must not be thought that the features of any one map at this scale will remain unchanged; the policy of developing editions will also no doubt include modifications to map style to cope with changing demands.

OTHER GOVERNMENT MAP-PRODUCING AGENCIES

Map series are produced by a number of other official bodies, although the actual map printing is frequently carried out by the Ordnance Survey. These bodies include the following:

a Ministry of Defence
    1 Military Survey – maps at various scales
    2 Hydrographic Department – marine charts at various scales
    3 Air editions of military maps or ICAO charts for air use – various scales
b Department of the Environment: produces maps at various scales, chiefly for planning purposes. Thematic maps include the maps of the Desk Atlas of Great Britain (about Half Million). A number of maps at a scale of 1/625,000 covering Great Britain in two sheets have been prepared by the Ministry, the Ordnance Survey and other bodies and

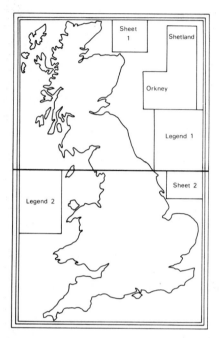

Figure 5.1   Northern and southern sheets of the Ordnance Survey 1/625,000 map of Great Britain, showing layout

form a series practically amounting to a national atlas in their coverage (Figure 5.1).

c Ministry of Agriculture, Fisheries and Food: agricultural land classi-fication maps, 1/63,360 and smaller scales

d Soil Survey of England and Wales: maps in a series designed to cover the whole country at the scale 1/63,360 together with special sheets, e.g. 1/25,000 Romney Marsh

e Institute of Geological Sciences – Geological Survey: maps at various scales from 1/25,000 for special areas to 25 miles to the inch. The major series is the 1/63,360 'New Series' covering Great Britain on the sheet-lines of the Ordnance Survey Third edition (Small sheet series). Older maps showed geological outcrop boundaries in black and were hand coloured; current series maps are colour printed and the National Grid Reference System is overprinted on the Third edition base. New revisions use a more recent topographic base, e.g. Seventh series Ordnance map, or the 1/50,000 series.

f Maps of overseas territories are produced by the Department of Overseas Surveys and include topographic series, town plans and general maps for many former colonies and other countries in the Commonwealth.

LAND UTILISATION MAPS

Maps of land use in Great Britain have been produced in the past on the framework of the 1 inch to the mile map and are currently on the sheet-lines of the Ordnance Survey 1/25,000 (Second edition) covering a ground area of 10 kms north–south by 20 kms west–east. These are based on field surveys broadly carried out in the early 1960s on the 6 inch to the mile map and are published by the Second Land Utilisation survey of Britain (Miss A. Coleman, King's College London).

COMMERCIAL MAPPING

Few topographic map series are prepared by commercial and private organizations – the chief exception being a 1/100,000 map series produced by John Bartholomew & Sons Ltd, of Edinburgh. A variety of atlases are, however, produced by such bodies, for example, John Bartholomew & Sons Ltd (who also print, amongst others, *The Times Atlas*), Readers Digest Association, Clarendon Press ('Oxford Economic Atlases'), Wm. Collins, Sons and Co Ltd Collins and Longmans atlases and Thomas Nelson & Sons Ltd.

Road maps, wall maps and so on as well as atlases are produced by other organizations, for instance Geographia Ltd (road maps, thematic maps), A. W. Gattrell and Co. Ltd, the George Philip and Son Ltd group and the Automobile Association.

Further details of the types of publications of all these commercial organizations and of the official mapping agencies are contained in their catalogues, many of which are free on request, and a summary appears in the reports to the International Cartographic Association, which are published in the *International Yearbook of Cartography*.

6

Other map series

Before considering other map series it is necessary to define the various functions that any mapping body *may* undertake. These are:

    a Geodetic surveying: the accurate determination of surface position and of earth shape

    b Topographic surveying: the detailed surveying of the topography of the country by ground and air methods

    c Making cadastral records: maps, deeds and ownership records, land registration

    d The compilation of thematic maps of various types: e.g. ICAO charts, hydrographic charts, geological maps, land-use maps

    e Map production: the final preparation and printing of maps of all kinds, but especially of topographic series.

The British Ordnance Survey does not itself carry out all these functions (thus, although it makes cadastral maps, records of land ownership are the function of the Land Registry office). In Western Europe the above activities are even more dispersed amongst both national and even municipal organizations as well as commercial groups. Nevertheless, close co-operation exists between the various sections of mapping and surveying since each depends on the work of others. In most cases the basic survey systems use an independent grid related to some map projection of the earth's graticule and it may prove difficult to 'fit' adjacent maps across international boundaries. Additionally, each national survey uses an independent datum for height calculations; these are summarized for Western Europe in Table 6.1.

    Survey organization is frequently the responsibility of a military survey body or is associated with one (as has been the case with the Ordnance Survey). Cadastral surveys and land registration involve the preparation of maps at a large scale (e.g. 1/500). Maps of a similar scale are also required for planning purposes but are not necessarily needed in a printed and published form. In a number of cases this need can be met by having

Table 6.1 Table showing relationships between datum levels of some European national mapping organizations (in metres) (from Imhof 1965)

Country	Datum	Differences (metres)
Belgium	Brussels 1892	+ 2·33 = (Netherlands)
	Ostend	+ 2·31 = (Germany)
		+ 2·31 = (Netherlands)
		+ 2·29 = (Germany)
Denmark	Erritso	+ 0·09 = (Germany)
Finland	Helsinki	+ 0·08 = (Sweden)
France	Marseilles	+ 0·25 = (Germany)
		+ 0·19 = (Spain)
Germany (W)	Berlin	− 0·06 = (Switzerland)
		Denmark (see above)
		Belgium (see above)
Great Britain	Ordnance Datum	− 0·05 = (France)
Italy	Genoa	+ 0.25 = (Switzerland)
Netherlands	Amsterdam	Belgium (see above)
Norway	Tredge	+ 0·41 = Sweden
Portugal	Cascais	− 0·21 = Spain
Sweden	Stockholm	Norway (see above)
Switzerland		Italy (see above)
		Germany (see above)
Spain	Alicante	Portugal (see above)
		France (see above)

map originals on a stable film and making *diazo* copies as required (Chapter 13). Alternatively, various local authorities may produce their own maps under special arrangements or licence.

The general topographic map needs of the military authorities may be served by the production of only small and medium scale map series. Larger scale map series may then be produced by a different national organization or even by the major municipalities.

MAPS OF WESTERN EUROPE

In general, the main map scale at which there is considerable coverage of the countries of Western Europe and which is most readily available in a published form is 1/25,000, or 1/20,000. The major mapping bodies and map scales of various European countries will be briefly considered here.

Belgium

1 The official mapping body is the *Institut Géographique Militaire de Belgique* under the control of the Ministry of Defence. Geodetic and survey work is carried out by the *IGM*, leading to the publication of topographic map series at scales from 1/10,000 to 1/250,000. Thematic maps forming part of the national atlas and soil maps of Europe and Africa are produced by this organization, and also administrative maps of Belgium. The basic topographic map of Belgium is at the 1/25,000 scale, with a 1/10,000 map available for some areas. Rather generalized map series at 1/50,000 (*Type Rapide*), 1/100,000 and 1/250,000 scales are also available, but a new 1/50,000 topographic map series of Belgium is in progress. The 1/25,000 map is a six-coloured map with a multilingual legend (Flemish, French, English) and it portrays some aspects of rural land use as well as topographic detail, contours being at $2\frac{1}{2}$ m or 5 m intervals, depending on degree of relief.
2 Large scale mapping (1/10,000 and larger) or urban agglomerations in Belgium is the role of the *Service de Topographie et de Photogrammetrie* of the *Ministère des Travaux Publiques*. The city regions of Antwerp and Brussels are so covered.
3 Thematic map series in course of production in Belgium are a soil survey and representative vegetation maps (1/20,000 scale) and detailed geological and geomorphological maps (1/25,000 scale).
4 Commercial maps include road maps and town plans by the firm of *R. de Rouck* (Brussels).

Denmark

1 The Danish official mapping agency is the Royal Danish Institute of

Geodesy (*Kongelik Dansk Geodatisk Institut*). National map series are published at scales from 1/10,000 and 1/20,000 to 1/1 million – this last being in the style of the Internation Map of the World (see page 67); longitude is referred to the International meridian and to the Copenhagen meridian.

2 Larger scale maps are prepared by the Danish Directorate of Land Registry.

3 Thematic publications include the 1/160,000 and 1/100,000 series of geological maps produced by the Danish Geological Survey and an Atlas of Denmark published commercially by Hagerup. This is a joint production involving university geographers, the Royal Danish Geographical Society and other bodies.

France

1 The official topographic map series of France is published by the *Institut Géographique National* at scales of 1/20,000 (old series), 1/25,000 (new series), 1/50,000 and 1/250,000. Plastic relief maps (moulded) have been produced by the *IGN* at various scales from 1/50,000 to 1/1,750,000. Since the early 1970s the *IGN* has also published topographic maps designed for a general market under the title of 'tourist maps' at both the 1/100,000 and 1/250,000 scales to cover the whole country.

2 Larger scale maps are produced by other organizations such as the *Service du Cadastre* – the descendant of the cadastral survey initiated by Napoleon – whose principal interests are in land subdivisions and land holdings. Planning purposes are served by the *Division des Travaux Topographiques*. Both these organizations prepare plans at scales ranging from 1/500 to 1/5,000 although some smaller scale maps (down to 1/20,000) are also produced.

3 A number of other official mapping organizations exist in France of which the *Bureau de Recherches Géologiques et Minières* is probably the largest. Marine charts and aeronautical charts are respectively the responsibilities of the *Service Hydrographique de la Marine* and the *Service de l'Information Aéronautique*.

4 Studies of vegetation distribution, forestry and ecology are also served by published maps together with some detailed maps of geomorphology. Wine-producing areas are also covered by maps at various scales and these are published by several different organizations but form a distinctive map type in the cartography of France.

5 Commercial mapping in France includes the maps produced by the *Manufacture Française des Pneumatiques Michelin* – the well-established 'Michelin' maps produced for tourism at scales varying from 1/50,000 to Quarter Million. These maps are not confined to France alone: countries covered by some eighty maps (and guide books) are France, Germany,

Austria, the Benelux countries, Spain, Great Britain, Italy, Portugal, Switzerland and parts of Africa. Early maps carried mainly touring and road information but more recent editions of the 1/200,000 map series also show relief by spot heights and hill shading.

6 Atlases published in France include a wide range of national, regional, historical and economic themes.

Luxembourg

1 Production of topographic maps of the Grand Duchy is under the control of the *Administration du Cadastre et de la Topographie*. Maps at 1/25,000 scale are printed by the *IGN* (Paris) and conform to the sheet-lines and style of the French series. Current developments include the replacement of this scale by maps at 1/20,000 of similar style and on the same sheet-lines. The maps are in four colours and contours are at 5 m intervals. A 1/50,000 map series is also available.

2 Cadastral plans at 1/10,000 are produced by the Luxembourg Survey Office and copies can be obtained for administrative, planning or land registry purposes.

3 Geological maps at 1/25,000 form a complete series (1/50,000 for the Ardenne area) and are published by the *Service géologique de Luxembourg*.

The Netherlands

1 Map production in the Netherlands has a long history, and Dutch cartography and printing skills have resulted in many excellent mapping organizations, both governmental and commercial. Official maps are produced by the *Topographic Service* at Delft, at 1/10,000, 1/25,000, 1/50,000, 1/100,000 and 1/250,000 scales. The largest scale is only available in a grey outline but the remainder are at least six-colour productions. A comprehensive official catalogue is available and includes map extracts.

2 Other official map-producing bodies include the *Waterstaat* (Transport and Water Management), Cadastral Service, Geological and Soil Surveys and Hydrographic Bureau. Water control maps, soil and geological maps are all at 1/50,000 scale. Detail maps at a larger scale are produced for specialized needs, for instance maps of the Rivers Maas and Rhine at 1/2,000 and 1/5,000 for maintenance purposes. Smaller scale maps are also produced by the Geological Survey including a detailed map at 1/600,000 with text in Dutch and English.

3 Large scale maps (1/10,000, 1/5,000 and so on) are also prepared by the municipalities from their own as well as from the official surveys. These, together with thematic maps, planning maps and others, are available for the major urban areas – Amsterdam, Rotterdam, The Hague.

4 Thematic maps are well represented in the sheets of the national atlas

(basic scale 1/600,000), *Atlas van Nederland*, and the historical and agricultural atlases of the Netherlands. Commercial map production shows a specialization in reproduction of old maps and atlases, e.g. *Theatrum Orbis Terrarum*, reproducing atlases of the sixteenth, seventeenth and eighteenth centuries. Road maps are produced by the Royal Dutch Touring Club, the *Cartografisch Instituut*, Bootsma, and atlases by *Wolters – Noordhof, NV* and the *Elsevier* Publishing Company.

Norway

1 Maps at scales of 1/25,000 and smaller are produced by the official organizations – *Norges Geografiske Oppmåling* – one of the oldest European bodies of this kind, founded in 1773. The basic map scale is the 1/50,000 and longitudinal references are to the Oslo Meridian as well as to the international graticule.
2 Geological maps form a complete series at 1/250,000 scale and hydrographic surveys of home and overseas waters are made at various scales by other official organizations. Polar areas and Spitzbergen are mapped by the *Norsk Polarinstitutt*.
3 Large scale survey maps are prepared by some of the municipalities for planning and other purposes and a commercial map producer is *Cappelen* who publishes tourist and road maps at small scales (e.g. 1/325,000 and 1/400,000).

Spain

1 Although the fundamental topographic map of Spain is at the 1/50,000 scale, this requires about 1,000 sheets and is not fully available. Closely related sheet-lines at 1/800,000 1/400,000 and 1/200,000 of the other *Mapa Militar* series are available; all these maps are produced by the *Servicio Geografic del Ejercito*, Madrid.
2 Geological maps follow the same sheet-lines and are being produced by the *Instituto Geológico y Minero* at the 1/400,000 and 1/200,000 scale.
3 Tourist maps at scales of 1/200,000 have been produced for selected areas by Firestone-Hispania, Bilbao, and a smaller scale (1/500,000) road map series for the entire Iberian Peninsula is also published by this firm.
4 An official series of a national atlas began to appear in 1966 and is produced in separate parts rather than separate sheets. This is the responsibility of the *Instituto Geografico y Cadastral*.

Sweden

1 There are a number of important map-producing agencies in Sweden. The official Geographical Survey office (*Rikets allmänna kartwerk*) produces the topographic map on the basis of a Transverse Mercator grid

(central meridian 2° 15′ W of Stockholm) but the actual printing is carried out by either private or state printers on a contract basis. Map series include a photomap and the Economic Map of Sweden (at 1/10,000 and 1/20,000 scale) as well as the 1/50,000 topographic series (1/100,000 in mountainous areas). The Economic Map is based on the large scale photomap which it has generally superseded and is a four-colour ortho-photomap in which the background tones of the rectified air photographs are printed in different colours according to land use, e.g. yellow for arable land, planimetric detail in black and a green tone elsewhere. Contours are overprinted in brown.

2 Land survey requirements in Sweden are undertaken by the *Kungliga Lantmateristyrelsen* (National Land Survey Board). Property divisions are normally recorded on black editions of the 1/10,000 base map but larger scales (1/400 and 1/1,000) are also used. This organization is also involved in the compilation of a new Property Register in which map grid co-ordinates will be used to identify plots. This register will form part of the Swedish National Data Bank (Hägerstrand 1967).

3 Other official bodies are concerned with the production of marine charts, aeronautical charts, geological maps and so on, and another organization handles sales and distribution of official maps.

4 A large commercial organization – the *Esselte Aktiebolag* or Esselte map service – is one of the largest map-printing bodies and produces many of the official map series for the Geographical Survey office as well as tourist maps, atlases and town plans, for its associated publishing houses.

Switzerland

1 The national topographic maps of Switzerland are produced by the official organization, the *Service Topographique féderale*, at scales of 1/25,000, 1/50,000 and 1/100,000.

2 Larger scale maps (1/5,000 and 1/10,000) are the responsibility of the *Direction fedérale des Mensurations*. These form the basis of maps for planning and land registration requirements.

3 Thematic maps and maps of special areas are well represented in Swiss cartography, ranging from geological and vegetational small scale surveys to a large scale (1/10,000) map series of the Aletsch glacier. This latter shows many of the features common to Swiss map series, being prepared from aerial and ground surveys, with extremely fine colouring and print-ing details of the complex structure of this area. Contours are shown at 10 m intervals, brown or black to indicate relief shading on rocky areas and blue on the ice areas. Rock outcrops are drawn and lettering is very neatly executed. Hill-shading techniques, the use of pastel colours and light tones, have been developed by Swiss cartographers over many years to a very high standard.

4 Commercial mapping organizations based in Switzerland, e.g. the firms of *Kummerly & Frey, SA*, and *Hallwag, SA*, are well known throughout most of Europe for motoring maps as well as thematic maps, town plans and so forth.

5 The Swiss *Kartographisches Institut* at the Federal Polytechnic in Zürich has also been responsible for a number of cartographic productions and research in the subject. Probably the most widely known of its sponsored publications is the *Atlas of Switzerland*, edited by Professor E. Imhof, which uses the 1/500,000 map of Switzerland as its main base map.

Federal Republic of Germany

1 The main survey network for the Federal Republic of Germany is under the direction of a federal organization, the *Institut für Augewrandte Geodäsie*, which also produces smaller scale maps of the country (scales of 1/200,000 and smaller). Other federal organizations are the *Deutsches Hydrographisches Institut* (soil survey), *Bundesamstalt für Bodenforschung* and ministries concerned with federal planning and statistics. Again these are chiefly responsible for small scale publications or atlases.

2 Larger scale topographic map series and also thematic maps depicting soils, geology and so on are the work of the various states (*Länder*). Although prepared and published by the *Länder*, these maps conform to an overall sheet-line system and specifications so that there is available full topographic coverage for the country. Certain areas are only available in older monochrome styles at the largest scale (1/25,000) but generally this is a fully coloured map depicting contours, hill shading, communications, buildings and vegetation in conventional colours of brown, blue, black, red and green.

3 Much progress remains to be made with larger map scales but recent work on an orthophotomap at 1/5,000 (of Nordrhein–Westfalen) may enable gaps to be filled in extensively built-up areas. Contours on this map are shown by black overprinted lines, names and roads in white.

4 Commercial mapping organizations in West Germany show a strong concentration on atlas production, e.g. *Kartographisches Institut Bertilsmann, Georg Westermann Verlag, Verlag Herder KG*. Road maps are produced by the *Kurt Mair* organization and *Falk-Verlag* of Hamburg.

MAPS OF NORTH AMERICA

USA

1 The official topographic mapping organization for the USA is the United States Geological Survey (USGS), which also collects, analyses and distributes a variety of information on American land, mineral and water resources by means of a range of thematic maps, survey data, aerial and

space imagery at different scales. There is a national cartographic information centre (established in 1974) that collates the work of several federal agencies so that enquiries to the NCIC are then directed to the appropriate organizations holding the desired data.

2 Topographical maps are produced for the whole country at scales of 1/250,000 or smaller with sheet-lines based on the earth's graticule (Figure 1.9). Larger scales are available for many areas, at 1/125,000, 1/62,500 or 1/24,000 (see page 14) but such coverage is likely to be patchy and reference must be made to the state index diagram or other sources (Figure 1.10).

3 Many privately produced maps are to be obtained on a state or county basis and relatively inexpensive motoring maps are published on behalf of many of the major gasoline (petrol) companies. The basic mapping is based on the work of the federal agencies and the printing work may be carried out by large publishing houses such as Rand McNally.

4 Although only some of its publications are actually of the USA, the map production work of the National Geographic Society, Washington, DC, must be mentioned here. In addition to the well-known 'yellow' magazine this organization publishes several maps each year which are the result of much research and compilation and have a distinctive style (Figure 6.1).

Figure 6.1 Extract from National Geographic map ('Close-up USA': Illinois, Indiana, Ohio, Kentucky) of part of the state of Ohio to illustrate the topography and style of this distinctive map series. (The original is in colour and therefore this figure has lost some of the character of the map.)

This is somewhere between that of topographic, touristic and thematic and is characterized by its lettering, careful use of colour and relief presentation.

Canada

1 The national mapping organization of Canada is the Department of Mines and Technical Surveys which was established as recently as 1943 and which in 1951 established a geographical bureau responsible for collecting and organizing geographical resource information. The national topographic map series include a mixture of scales ranging from 1/50,000 to 1/506,880 of both imperial and metric units but they are closely related in sheet-lines linked with the Universal Transverse Mercator grid. Only the smallest scale is complete for the whole of Canada.
2 Much earlier mapping work was associated with the development of the country, particularly in the pioneer areas of the north: the Geological Survey was established in 1841. Photographic surveys of various types were established at an early date and the whole country has been covered by them.
3 Thematic mapping in addition to geological maps include land-use maps, tourist maps of special areas, soil maps and economic survey maps, all produced by various government departments.

MAPS OF AUSTRALASIA

Australia

1 The national mapping organization for Australia is a federal body, the Division of National Mapping (NATMAP), which primarily produces maps at the 1/100,000 scale and smaller.
2 Individual states of Australia have map-producing organizations, variously known as Departments of Lands and Surveys (Central Mapping Authority in the case of New South Wales), and these are responsible for cadastral and larger scale topographic mapping than that undertaken by the federal body.
3 The equivalent of a national atlas is in map series form and is entitled the 'Atlas of Australian Resources' (Second series). This consists of some thirty coloured maps which can be obtained separately or as a complete atlas; each map sheet is accompanied by a booklet giving a commentary on the data shown on the map.

New Zealand

1 The national organization responsible for the official topographical maps of New Zealand is the Department of Lands and Survey and the standard

map scales are at 1/25,000, 1/63,360, 1/250,000 and 1/500,000. These are based on a Transverse Mercator Projection with a grid dimensioned in yards and, because of the arrangement of the two islands that comprise the country, there are two standard meridians.

2 A wide range of thematic mapping is directed chiefly at the resources of New Zealand so that, in addition to the geological maps, there are various soil and vegetation mapping representatives. Mostly these are at small scales (1/1 million or 1/2 million) but some regions have been mapped at larger scales.

MAPS OF SOUTH AFRICA

1 The official mapping organization for the Union of South Africa is the Trigonometrical Survey Office, which was established in 1919. The basic map series is at 1/50,000 but the topographic map types range from 1/12,000 to 1/70,000 with smaller scales down to 1/2½ million also published. Most of these maps are bilingual productions in English and Afrikaans.

2 The basic geological mapping of the country is at the 1/250,000 scale and is the responsibility of the Geological Survey with its head office at Pretoria.

3 Thematic maps have been prepared nationally by government agencies including the Natural Resources Development Council and there are in addition many commercially produced maps for tourist purposes. The petrol companies in particular have sponsored road maps somewhat as in North America.

7

Atlases and world map series

An atlas is defined as 'a collection of maps designed to be kept (bound or loose) in a volume' (Royal Society 1966). The term 'atlas' generally invokes a somewhat more limited image – that of the fully bound reference volume containing maps of most countries of the world. On the other hand, many specialized atlases have been produced depicting features of a limited area. Atlases can perhaps be conveniently classified as:

1 Library and reference atlases
2 School atlases
3 National atlases
4 Regional atlases
5 Thematic atlases.

Common features of the first two categories are that they include collections of maps of both topographic and thematic types covering different countries, regions, continents and the world. The amount of detail shown on such maps depends on the purpose of the atlas, whether for junior school, secondary school or the gazetteer and reference function of the library atlas. The selection of areas covered in the atlas also reflects the purpose of the publication. The 'concentric' approach may be followed in school atlases with special neighbourhood maps and aerial photographs of the local region forming the introductory section, followed by comprehensive topographic and thematic map coverage of the homeland and less detailed maps covering a wider area such as the continental area and the rest of the world.

Atlases produced in any particular country tend to contain map coverage of that country to a greater depth than 'foreign' areas. This may be obtained by including maps at a larger scale or by selecting and stressing particular detail. Nevertheless the common feature of both scholastic and library atlases is their firm binding (as opposed to a loose-leaf format) and their inclusion of an index or gazetteer. This generally uses the conventional system of geographical co-ordinates of latitude and longitude but other methods are also used, for example, separate grid systems of numbers and letters for each page, or the system of time zones (cf. *Bartholomews Advanced Atlas*).

As a form of national inventory the atlas of an individual country is a valuable document but may sometimes have a prestigious aim. Since a full and adequate documentation in map form takes time to prepare and individual map sheets can quickly become outdated or need revision or replacement, it is normal to produce national atlases in a loose leaf form. These may then be mounted in a folder or contained in a wallet or box.

National atlases usually aim to present all possible distributional data for the country concerned and can be very extensive in their coverage. An introductory section will probably include maps of such features as relief, drainage, hydrographic regimes, soil types and climatic records. Resources and economic data – industrial and agricultural production, employment, trade, communications and similar information – will form another section. Demographic facts usually yield a further set of maps and special characteristics of the country may give rise to specialized maps, e.g. water and flood control features in the Netherlands, avalanche hazards in Alpine regions, vegetation patterns in tropical countries.

Table 7.1   Likely contents of a national atlas

Cartography and Topography	Settlements
Geology, Geophysics, Mineral Resources	Anthropology and Language
	Population and Demography
Geomorphology	Social and Economic Structure
Soils	Land Utilization and Agriculture
Climate	Public Utilities
Biogeography	Manufacturing Industries
Water Management	Transport and Commerce
Historical Geography (including Archaeology and Prehistory)	

The preparation of regional atlases has often followed the collection of data needed for planning purposes. In many respects the maps produced for these purposes will have much in common with national atlases and background information, historical, topographic and other details will provide an introductory set of maps. Other sections will reflect both the region and the purpose for which the atlas has been produced which may be purely for reference purposes or it may be as a basis for planning or, again, to set out planning proposals.

There are many thematic atlases in the specialist topics of both physical and human geography, e.g. climatological data for the former and urban for the latter. Spatial aspects of interest to the historian may also be summarized in map form and there is no lack of such atlases for many of the countries of Western Europe.

Within these main groups of atlases large variations occur in the types of map. The common factor is the presence of at least some topographic maps but by far the greatest number of maps in atlases of the national and regional type will be of the thematic form, both to illustrate distributions and to present plans and concepts. Large scale maps and plans will probably make up a major part of planning atlases with medium scale maps characterizing regional atlases and smaller scales used for national atlases. If a considerable amount of relevant and up-to-date data is to be presented this can be done by using methods of automated cartography, for example computer printouts, as in the Birmingham and West Midlands atlas (see Figure 4.5, page 49, for an example of lineprinter mapping with added information, main roads, outer local authority boundary).

INTERNATIONAL MAP SERIES

Although international co-operation on a proposed world map series began with its proposal at the 1 to 1 million scale at the 1891 meeting of the International Geographical Congress, progress was inevitably hampered by the wars and economic disturbances of the succeeding decades. Nevertheless, the map series was launched at the Paris Conference in 1913 when decisions were made as to the projection – a modified Polyconic for most of the earth's surface – and the methods of depicting relief, hydrology, populous areas and routeways. Work continued by the various member countries of the International Geographical Union and by 1949 approximately one-fifth of the map sheets had been published (405 out of a total in excess of 2,000). About half of these were to the full specifications of the International Map of the World (IMW), others being compiled by member countries with the same pattern, style and sheet-lines. Since 1954 the General Secretariat of the IMW has been part of the Cartographic Office of the United Nations and discussions have also continued as to the comparative roles of the IMW and the ICAO Aeronautical Chart at the same scale.

In 1962, at the Bonn Conference on the IMW, alterations were made to the specifications and sheet-lines, and the general objectives of the map were agreed as follows:

a To provide, by means of a general purpose map, a document that enables a comprehensive study of the world to be made for pre-investment survey and economic planning and also to satisfy the diverse needs of specialists in many sciences.

b To provide a base map from which sets of thematic maps can be prepared (e.g. population, geology, vegetation, administrative limits, statistical evaluation).

Figure 7.1 Sheet-lines of the world map at 1/2,500,000 scale, based on a rectangular world grid, to show how these sheet-lines are related to the world graticule

It must be noted that vegetation cover is not included on the base map but particular encouragement was to be made to complete such maps for the developing countries.

The ICAO World Aeronautical Chart at the 1/1 million scale was developed out of a United States Air Force series at this scale and its completion has been given priority by many individual countries.

A further map series at the 1/2·5 million scale initiated by the national cartographic office of Hungary should also be mentioned. This is the *Karta Mira* which began publication in 1964 and will eventually cover the world in 244 sheets (Figure 7.1). The official languages of the map are Russian and English, and the purpose of the map series is to meet the need for a world base map of a general geographical nature.

At a larger scale a further map series might also be considered to be international in scope. This is the Quarter Million (1/250,000) produced by many national survey organizations in both a Ground (or Land Map) and Aeronautical (or Air Chart) version. World coverage at this scale is far from complete but is summarized in Figure 7.2. Individual sheets may vary from purely outline in style to a coloured topographic map. In both New Zealand and Australia complete cover at this scale is provided in atlas form by the official mapping organization (Lands and Surveys, Wellington) for one country, and a private publisher (Readers Digest Pty) for the other. At one time the equivalent map of Great Britain (¼ inch to the mile) was produced in a loose-leaf atlas format but the present 1/250,000 series of maps is now aimed at the motoring public and is referred to as the Routemaster Series of nine sheets to cover England, Scotland and Wales.

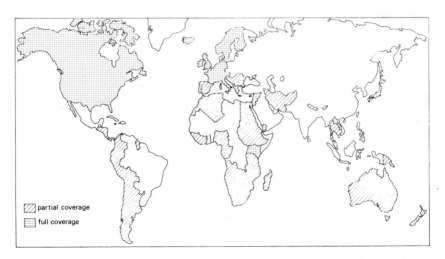

Figure 7.2 The world: coverage of topographic maps at the Quarter Million (1/250,000) scale

8

Appreciation and communication

CARTOGRAPHIC APPRECIATION

Maps are frequently so much an integral part of modern existence that they are taken for granted: the commonplace map in a familiar style becomes the norm and its value remains unquestioned. Faced with a new map type the user is tempted to make quick decisions of a subjective kind or may indeed use the map in an inaccurate and inadequate way. It is therefore necessary to be able to assess the value of any given map fairly, in as objective a manner as possible. This assessment is variously regarded as 'appreciation', 'critical appraisal' or 'critique' of the map. The chief dangers in making such an evaluation are those of subjectivity and of an intolerance bred from habitually using maps from a limited range of types. This short section puts forward both objective and subjective assessments which can be made of any individual map or series of maps.

Objective assessments: Such observations must necessarily be confined to dealing with precise details of the map under examination, namely

    a the overall features of the map, and
    b methods utilized for presenting the mapped data.

Subjective assessments: Under this heading come all those factors in which personal choice or preference enters, even if only to a limited extent. Such factors fall into the following groups:

    c suitability of mapping techniques
    d adequacy of detail depicted for map purpose
    e aesthetic considerations.

We shall consider each of these points in turn.

The overall features of the map

These include the size and format of the published map sheet(s) – including presence and size of margins – and the layout of the printed sheet (Figure 8.1), including these essentials:

title
sheet number
legend
scale(s)
additional data
date of survey
map production
reliability diagram
grid and its explanation.

At this stage in examining the map the assessor should merely record these various points in a purely quantitative form, for example, the size of the map in appropriate units, and the position of the other details listed above, noting their presence or absence.

Figure 8.1 Annotated diagram to show layout of typical Ordnance Survey 1/50,000 map and position of legend, scale, marginal information, grid and graticule numbering. Shaded area is that occupied by actual map information, outer line shows dimension of sheet of paper on which the map is printed.

Methods utilized on the map under consideration

Once again it is necessary to work through the various items shown on the map and to note the various methods used, which, if they are not standard or conventionally accepted methods, should be referenced adequately in the legend.

Thus it is essential to consider, on a topographic map, how the following features are indicated:

> Relief: contours, spot heights, shading, colour, other methods
> Drainage: distinctions between different forms, widths, etc., flow, water sources
> Coastlines: type, tide marks, foreshore
> Hydrographic features: symbols, contours
> Vegetation: shading, symbols, boundaries
> Land subdivisions: plot boundaries, fences, etc.
> Buildings and apparent generalization (will be indicated when considered in relation to boundaries which may, however, themselves be generalized)
> Settlements and their naming
> Communications: road, railway, canals, transmission lines
> Administrative divisions: boundary, symbols and naming
> Thematic data: isopleths or choropleths

The adequacy or otherwise of the legend can be considered here, and omissions, errors or departures from accepted practice, as these are clearly objective observations.

Suitability of mapping techniques

It is frequently difficult to assess this point without some knowledge of either the type of landscape depicted on the map, or without having attempted some precise investigation from the map. The latter course is always to be recommended because the map is a tool to be used and without having experience of the tool its value must remain in doubt. For example, a topographic map can be examined with a view to isolating the major relief units or the components of the transport network. Similarly a thematic map should be studied in order to extract information relating to a random choice of points. The ease of these various operations will provide a clear measure of the map's suitability (and also the adequacy of the legend or other explanatory text). The degree of subjectivity of the map assessment can thus be somewhat lessened. Nevertheless, much of this part of the exercise will necessarily be confined to an examination of the suitability of methods (see above, *Methods utilized on the map under consideration*).

Adequacy of detail depicted for map's purpose

While studying the mapping techniques used it is quite likely that this next point of adequacy will also have been considered. The essential point, however, is the *purpose* of the map. The suitability of the techniques (above) will merely indicate whether the map can be interpreted as a model of the landscape. Under this heading the whole validity and *raison d'être* of the map must now be examined. It is often pertinent at this stage to examine also the relations of the map sheets to one another and to look again at the objective points recorded earlier, for example map size, layout, additional or marginal data.

Aesthetic considerations

This section must embrace a wide range of topics and although the aesthetic and artistic features of the map should be considered there are other related points to be examined first. The general handling ability of the map (which stems from overall layout, suitability and so forth) and its overall suitability having been studied it remains to be seen if the presentation is attractive. Many maps are prepared on a commercial basis and the overall impression that they give to the user is a vital selling point. They must also contain the information the user wants, and be shown to contain this information. Methods of folding, covering and general salesmanship are not unimportant here.

CARTOGRAPHIC COMMUNICATION

As will be discussed in the next section, maps are made to be used. They are constructed for a whole variety of purposes but their overriding feature is the portrayal of some aspect of the earth's landscape to the map-reader. They are therefore a means of communication which at its simplest can be thought of in two stages:

1 The 'real' landscape is translated onto a map by the map maker
2 The map-user translates from the map to obtain a mental picture of the landscape.

Or it can be thought of diagrammatically, as shown in Figure 8.2, where the arrows represent the message or 'code'.

It is clear that some relevant information is likely to be lost at both the 'encoding' and 'decoding' stages in the transmission of the map message. Thus in Figure 8.2 the square for the map is less in area than the circle representing the real world and the triangle for the map-user's perception is smaller than the map square. However, it is quite likely that the map-user will be able to draw on other knowledge, so that he may incorporate

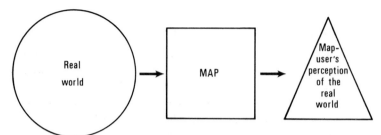

Figure 8.2   The map as a means of communication

additional data to that obtained from the map, hence the triangle symbol might be thought of as overlapping the square.

The terms 'map-user' and 'map-reader' are used somewhat indiscriminately, but it is possible to distinguish between them. By map-user can be meant the person who is merely utilizing a map to extract some specific information, for instance the location of a street, building, place or route. The map-reader, on the other hand, may be expected to extract rather more from the map, for example to be looking for some interpretation, explanation or relationship connecting features depicted on the map.

An analogy may be drawn between mere recognition of letters to make words in reading and the next step, that of understanding the message conveyed by a phrase or sentence. The analogy can be taken further so that it is possible to establish a dictionary of map symbols and then to examine their interrelationships as a form of grammar and syntax.

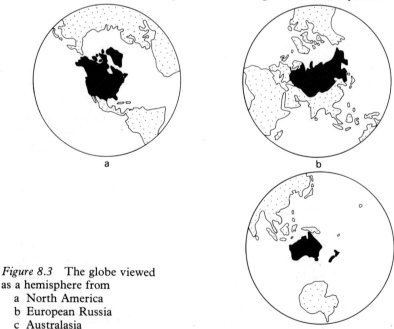

Figure 8.3   The globe viewed
as a hemisphere from
  a North America
  b European Russia
  c Australasia

At its lowest level the map dictionary will start with the map legend but this will need expansion to indicate possible arrangements and combinations that give rise to landscape feature. Thus contour lines and drainage can be translated into the physical features of hills and valleys.

It is of course possible to design maps in such a way that the message conveyed can be distorted. This may be done deliberately, perhaps for propaganda purposes, or it may be done in order to bring out particular points such as the spatial aspects of a perceived idea. This latter may be referred to as 'mental mapping' and such imaginative maps have built-in

Figure 8.4 Flanders: Mercator map extract (1562)

distortions due to experience or prejudice. The map-reader, too, accepts a number of long-established conventions such as those of layout, colour and orientation. Map scale is generally at the bottom, roads are frequently coloured red, water information is in blue and the north is at the top of the map. With global maps the standard conceptual layout of the earth also follows a standard pattern in which the equator is a horizontal central line and, because of the limitations of map projections, there is inevitably an

Figure 8.5 How the southern English view the United Kingdom

Figure 8.6   How Bostonians see the USA

outer line to the map, whereas the global sphere has a continuous surface.

Some examples of such maps may best be used to illustrate the above points and Figures 8.3 to 8.6 should be examined in detail. Figure 8.3 contrasts the viewpoint of global affairs when world relationships are viewed from North America, the USSR and Australia. An early map in which different scales were used for the nearer topography to that of the more distant parts is that of Flanders, drawn by Mercator in 1562. An extract from this is shown in a much reduced form in Figure 8.4 where the detail of part of Bruges is at a large scale with the rural countryside in the distance at a smaller one. The southern Englishman's view of Great Britain as seen from suburban London is well shown by the distorted map in Figure 8.5, while a similarly distorted view of the United States is that of the Bostonian as depicted in Figure 8.6.

9
Measurements
from maps

This chapter will consider the general problem of extracting quantitative data from maps; it is concerned with various precise measurements that can be made and also how these measurements may be used in attempting map analyses of the landscape.

Detailed topographical information is frequently required by geographers, planners and others, but it must be remembered that errors can occur whilst the data is being obtained from the map.

These errors include:

Scale errors: These can be due to variations in the size of the paper and can be reduced by always referring map measurements to the printed scale bars on the surface of the maps. These will change their shape in the same way as the features on the map itself. Such errors can be either positive or negative.

Projection errors and curvature of earth's surface: On small scale maps errors are introduced in representing the curved surface of the earth on a flat sheet of paper. Errors due to the type of map projection used are negligible on medium and large scale maps. They can be calculated by referring to the type of projection graticule used. (See standard textbooks on map projections.)

Slope and height errors: All information on the map is presented as if on a horizontal plane, usually that of mean sea-level; it is not presented as on the natural undulating surface of the landscape. If extreme accuracy is required allowance must be made for the difference between these two surfaces or alternatively it must be made clear if distance, areas and so on refer to the complex natural surface or to the more usually accepted sea-level plane. These errors are such that the mapped information is smaller than the true value (Figure 9.1).

Generalization errors: As has been indicated in Chapter 1, some degree of generalization is required between the details of the landscape and its representation on the map (Figure 9.2). Such generalization has the effect of smoothing out irregularities. Under this heading can be included errors in the original draughting of the map. Distances measured may therefore be somewhat too small compared with ground distances but, on the other hand, areas can be too great. A check can be made, by summing areas to compare with a known value, such as that of an administrative area.

Measurement or instrumental errors: Such are either gross errors due to carelessness or mistakes, or systematic errors resulting from the method used. In all cases measurements to printed lines on the map should be made to the centre of the line but many discrepancies can occur; for example when measurements are made to a printed road symbol – which may well be considerably enlarged from its true dimension – these might be taken to the centre of the colourfill or they may be made to the appropriate black lines on either side of the colour. Where a measuring instrument is being used, such as the planimeter for areas (see below), then a recognized 'drill' must be followed to reduce errors in the instrument.

Figure 9.1 Illustration of the fact that any horizontal projection will cover a smaller area than the true ground surface

Figure 9.2 Shape A shows the precise shape of a contour that becomes generalized to shape B for reproduction at a smaller scale. This generalization tends to be greater in area than the original.

It must be stressed that all measurements made from maps are subject to errors of the above nature and that although allowance can be made for them, it is meaningless to perform further calculations with these measurements to levels beyond their accuracy or significance.

TYPES OF MEASUREMENT

The main types of measurement made from maps are of distance, direction and area. *Distance* measurement clearly involves assessing lengths on the surface of the map – either against each other for comparison purposes, or against the known scale to obtain absolute values. Direct or straight-line distances may be required and present little difficulty as they can be taken directly with a pair of dividers or a scale placed on the map surface. More usually, distances are required along erratic paths: sometimes these can

NORTH POINTS

Difference of true north from grid north at sheet corners

N W corner	N E corner	S W corner	S E corner
1° 03′ E	0° 33′ E	1° 02′ E	0° 32′ E

Magnetic north about 7½° W of grid north in 1979 decreasing by about ½° in three years

Diagrammatic only

Figure 9.3   Some typical North points (grid, true and magnetic) from an Ordnance Survey 1/50,000 map, with accompanying notes to indicate more exact differences between these directions

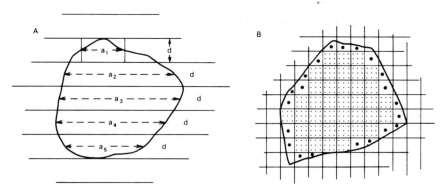

Figure 9.4   Area measurement methods

    A A series of parallel lines, equally spaced at distance d, cross the area to be measured. The distance between the shape's outer boundaries midway between each line is then measured (a_1, a_2, a_3, etc.). Each strip can then be considered as a rectangle of area d.a_1, d.a_2, etc., and these are summed. Alternatively, vertical lines are drawn on the 'give-and-take' principle, to omit an estimated section equal to an outer area that is included in each rectangle. The area is given by d(a_1 + a_2 + a_3 ...).

    B The area to be measured is covered by a regular pattern of squares (e.g. by placing a sheet of graph paper beneath the map on an illuminated tracing table, or by tracing the area onto squared paper). The total number of whole squares is summed, together with all partial squares. Area = sum of whole squares + half total of part squares.

be broken down into sections and each part taken with dividers, or it may be necessary to follow the active path with a map measuring wheel (*opisometer*). Any splitting of the distance into parts will introduce possible cumulative errors, either small and systematic or large and accidental (Maling 1968).

The measurement of *direction* is less often required but produces more problems than either distance or area calculations. The problem here is that of datum direction – usually that of north. Well-produced maps will generally indicate at least the direction of true north. Topographic sheets will also add that of magnetic north – for the benefit of field map-users – and probably a note as to the relationship between grid north and true north – a relationship that will vary from sheet to sheet (Figure 9.3).

Area measurement is usually carried out on the surface of the map and assumes that the discrepancy between this sea-level plane and the undulating natural surface is smaller than the errors of making the measurements. There are numerous ways in which areas can be calculated (Figure 9.4) and summaries appear in most standard textbooks. Frolov and Maling (1969) provide a discussion of the relative accuracies of these methods. The most usually available mechanical method is that of the fixed area polar *planimeter*. In this instrument a measure is made of the movement of a rod whose locus is constrained by having one end fixed to a radial arc (Figure 9.5).

The area to be measured is traced along its perimeter in a clockwise

Figure 9.5 The fixed-arm polar planimeter: this measures area by tracing the outline of the closed area under consideration in a clockwise direction and returning exactly to the starting point. The movement of a wheel at right angles to the planimeter arm is recorded in terms of whole revolutions and decimals of a revolution. The area swept out is given by this figure multiplied by the constant of the instrument. Since very small movements are measured, a vernier reading of the drum is necessary and it is also necessary to repeat the measurement, say, three times, and to take the mean of these three provided they show reasonable agreement.

direction with an index mark, starting from one convenient point to which the index must exactly return. Reading of the recording dials before and after so tracing the area's perimeter will give a value in instrumental units. These are in terms of complete and decimal fractions of revolutions of the wheel attached to the planimeter rod. They must be multiplied by some constant for the particular instrument to convert into areas in square inches or square centimetres. In order to obviate gross errors, at least three readings of each area should be taken and the mean calculated, provided there is reasonable agreement between the three readings. Values showing wide disagreement should be disregarded and the readings taken again.

Non-instrumental methods of measuring area are also possible. Of these the most simple is probably that of transferring the required area to a grid of printed squares either by tracing off or by superimposing transparent printed tracing paper (e.g. with 1/10 in squares or 2 mm grid). The number of whole squares (W) and partial squares (P) are separately summed and the area is approximately given by

$$A = W + \frac{P}{2}$$

that is, the total of all complete squares plus half the total of partial squares. Alternatively a visual assessment may be made so that areas excluded in some partially occupied squares are balanced by areas included elsewhere (Figure 9.4B).

Figure 9.6   Area calculation by ordinate measurement: if a regular series of vertical ordinates are measured to the irregular boundary of the area from a straight line, either of two formulae may be used in order to calculate the area:

a Simpson's rule

$$\text{Area} = \frac{d}{3} (a_1 + 4a_2 + 2a_3 + 4a_4 + \ldots + 2a_{n-2} + 4a_{n-1} + a_n)$$

b Trapezoidal rule

$$\text{Area} = d \left(\frac{a_1}{2} + a_2 + a_3 + \ldots + a_{n-1} + \frac{a_n}{2} \right)$$

where d is the common distance apart of the ordinates a_1, a_2, etc. Note that an ordinate must be measured at both the extreme ends of the lines. If the outline approaches the horizontal axis, this ordinate will become 0, but it must be included in the formulae. With Simpson's rule it is necessary that there are an even number of divisions (i.e. the number of ordinates must be odd).

This leads to another method of area measurement in which the area is covered by a series of parallel, equally spaced lines (Figure 9.4A). Bounding lines at each end of the parallels can be drawn on this 'give-and-take' method but almost certainly an irregular shape will remain as in Figure 9.6. This is then measured by other methods, e.g. Simpson's Rule (see below). Geometrical calculations of area (mensuration methods) provide an absolute value and are possible in cases where regular outlines are considered – rectangles, parallelograms, triangles, sectors of circles and so on.

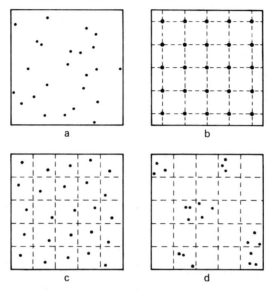

Figure 9.7 Point-sampling patterns: sampling methods may be used in order to choose locations for further study instead of undertaking a tedious examination of a complete area. In area-measurement usage, sampling over a whole region will consist of counting areal activities at the sample points and using the number of occurrences divided by the total number of points as an indication of the proportion that activity bears to the total area.
  a Random pattern
  b Regular, or systematic, pattern
  c Systematic unaligned random pattern
  d Random pattern within specified grid squares (nested random sampling)

Where it is possible to reduce the shape to an irregular outline on one side and a straight line on the other, standard mathematical methods of area calculation can be used – either Simpson's Rule or the Trapezoidal Rule (Figure 9.6). This will require the measurement of a series of regularly spaced ordinates at right angles to the straight side.

Area measurements involving a 'sampling' technique may also be used: these have been described by Haggett (1965) and Frolov and Maling (1969) and have been used in conjunction with the Second Land Utilisation Survey of Britain. They yield a high accuracy, coupled with speed of operation, and require no measurement of areas as such but only the counting of the number of occurrences of the data under examination over the study area, map sheet or series of maps. Various sampling patterns are illustrated in Figure 9.7, and these are normally marked on a transparent overlay which is then placed over the data map. A count is then made of the activity, e.g. land use, at each point, and these are summed for the area or map sheet. If the area extends over several map sheets care must be taken neither to exclude the join of the sheets nor to count it twice.

Angular measurements on maps to ascertain aspect, bearings or other criteria best quoted in degrees or radians must be made with caution. Large

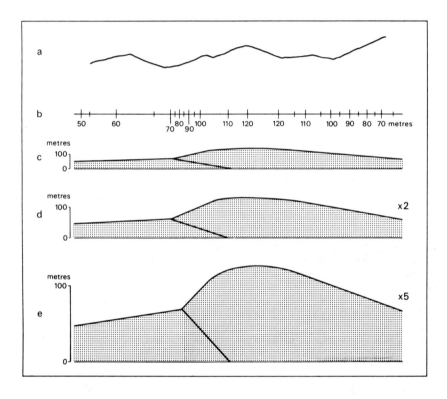

Figure 9.8 Profiles: simple profiles are drawn along either straight lines or, more usefully in analysis, along watersheds, river courses, or at right-angles to successive contours to show maximum slopes. The plan view may thus be as in (a) but this must be reduced to a straight line (b) before construction of the profile as a simple line graph (c). Horizontal and vertical scales need not be the same; a vertical exaggeration can help stress relief or slope changes: (d) and (e) are the same profile with times two and times five vertical exaggerations. However, if geological dip lines are inserted, shown by sloping lines, the effect of vertical scale exaggeration will be to increase the angle of dip to a ridiculous extent.

scale topographic map series, however, are generally on an orthomorphic projection and thus angular properties on the ground and the map are identical. The only problem remaining is that of a datum or starting point for bearings. True north, magnetic north and grid north should be quoted in the map margins or the relationship between some of these directions and lines printed on the map indicated. Angles and bearings on non-orthomorphic projection smaller scale maps cannot accurately be measured and only general statements can be made in relation to cardinal compass directions. Because angles are frequently measured on a whole circle basis, a 360° protractor should be used.

Measurement can be used in map analyses in many different forms, either to produce new diagrams that will themselves be analysed, or to make a statistical analysis of the extracted data.

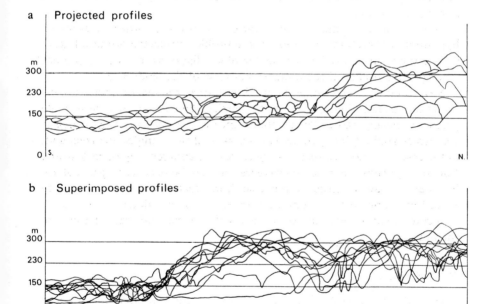

Figure 9.9 Profiles used in preliminary morphological investigations: a series of regularly spaced parallel profiles may be combined to search for possible regional patterns as in (b), which shows *superimposed profiles*. The effect of a panorama is produced by the method of *projected profiles* (a), in which only those sections are drawn that are not obscured by nearer profiles. These profiles are drawn along south–north gridlines on the southern edges of the Pennines (from Clayton 1953: 30, Figure 3).

DIAGRAMS PRODUCED FROM MAP MEASUREMENTS

Under this heading can be included such studies as the construction of profiles, block diagrams and summary diagrams such as network patterns, and other 'models' of the landscape based on the map evidence.

Profiles

The method of obtaining a 'side elevation' view of the land form is shown in Figure 9.8. Spot heights and contours provide a series of discontinuous values for height which can be plotted against distance along the selected line of profile. The latter can be a straight line between two points, a grid line, or can be a sinuous line such as a watershed, road or stream line. The resulting single profile can be made to yield more clues as to the physical landscape if it is then compared with the geological map so that underlying rock information can be added, and the profile is converted into a cross-section.

Profiles can be combined in a variety of ways in order to establish hypotheses of landform evolution (serial profiles, projected profiles; Figure 9.9) or they can be used as a basis for block diagrams in order to present a perspective pseudo three-dimensional view of the landscape.

Nevertheless, profiles must be used with care, the most important consideration usually being that of the vertical dimension. A true-to-scale profile is usually rather flat in appearance and a *vertical exaggeration* is frequently applied. This immediately makes all slope angles incorrect and, in the case of geological sections, increases the apparent dip of rock strata. For some purposes a massive exaggeration may be necessary, particularly in order to draw attention to relict landform features, but it is unwise to exceed five times the horizontal distance: a common dimension is that of 1 inch to 1,000 ft vertically, combined with a horizontal scale of 1 inch to the mile.

Block diagrams

The need for a pseudo three-dimensional diagram is often encountered in the interpretation of landforms, but is not confined to studies of the physical landforms since population and agricultural production figures, for example, may also be used as the 'z' component (assuming the two-dimensional map is in the form of x and y directions and that the third dimension of height or other function is z). Various types of block diagram have been developed (Monkhouse and Wilkinson 1971) but the most conveniently constructed is the simple isometric diagram prepared from a series of profiles or from transformed contour patterns (Figures 9.10 and 9.11). In addition, a mechanical method of drawing block diagrams is

a

b

c

30° 30°

d

Figure 9.10 Block diagrams

a A perspective block diagram, in which the sides and detail will appear to converge towards two vanishing points on the perspective horizon

b A simple isometric block diagram, one side inclined

c An isometric block in which both sides are inclined

d Extract of a simple contour pattern covered by a regular grid as a basis for the construction of block diagrams (e) and (f)

e Construction of a block diagram by reconstruction of the contour pattern in a series of 'layers': the detail of each of the squares in (d) is transferred to the isometric grid at the appropriate level. This is the basic method of the *Perspektomat*, an instrument that aids in the drawing of block diagrams.

f Construction of a block diagram by drawing successive profiles across the terrain (some are indicated in broken lines) and finally drawing hachures or shading down the slopes

e

f

available in the *Perspektomat* – a device that incorporates a pantograph mechanism so that not only is the appropriate viewing angle transmitted to the diagram but also changes of scale are possible.

Slopes

Although it is easy to measure slope angles from a contoured map by scaling off the distance between two successive contour lines, the student of land-forms is sometimes concerned to obtain 'regional' values of slope or 'average' slope over a wide area. On some maps, notably those of France and Germany, the calculation of hillside slope is made easier by the inclusion of an additional scale. A number of complex methods have been devised for estimating average slope and these are described in Monkhouse and Wilkinson (1971).

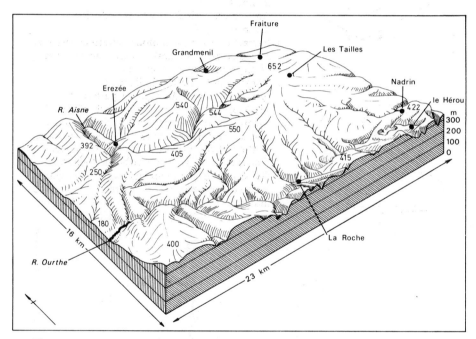

Figure 9.11 Annotated block diagram of part of the Belgian Ardenne using 'hachuring' symbols but originally constructed with the aid of the *Perspektomat*

Other diagrams

Much of the data extracted from measurements will result in tables of data. This may then need representation in a cartographic form or, more normally, by some statistical diagram – a graph of some kind (see p. 95). Simple linear diagrams to indicate networks, hierarchies or other relationships may

Figure 9.12 Numerical data on maps: New Zealand, South Island, with statistics illustrating various aspects of human geography
  a Population gains by migration, 1956–61
  b Distance in kilometres between main centres
  c Factory production, 1967–8 (all industries)
  d Electric power stations, number of sets

also be drawn and dimensioned by writing the appropriate values, such as distances, on the diagram (Figure 9.12).

Data relating to individual points can be presented in a map form merely by indicating the point (e.g. by a dot or cross symbol) and entering the value of the data alongside.

It is possible, however, that values for the data can be expected to occur for positions for which no precise reading is available. In other words, data values must be interpolated, provided it is known that the phenomena so recorded has a continuing occurrence over the whole area. If the phenomena is known to be of isolated and separate occurrences then no continuum can exist and any interpolation is purely hypothetical.

The interpolation of values between known values forms a way in which isolines can be constructed; or, conversely, from a map of isolines it is possible to suggest the value of a particular phenomenon at any given point. Interpolation or extrapolation of data from simple graphical figures requires an understanding of the way in which the graph is behaving – or the relationship between the graphical data. This relationship can be approximated within statistically calculated limits known as 'confidence limits' and the behaviour of the graph or functional relationship between the variables is expressed by the 'regression'. In the case of interpolation of isolines or the extraction of data from isoline maps it must be assumed that there is a constant change of data between the isolines or in all directions from a given value (Figure 3.9).

It is possible for many of the operations referred to above to be carried out by a computer program, providing that the initial map data is already available in co-ordinate form (that is, Cartesian co-ordinates and value of data). The tedious calculation of rates of change between adjacent points are then performed by the computer and the position of the isolines obtained.

More sophisticated analyses of data can be made by extensions of the methods of interpolation and/or calculations of regression values mentioned above. Such analyses may be required in order to examine the three-dimensional properties of mapped data and their functional relationships in three dimensions. In these cases the techniques of study are referred to as a study of 'trend surfaces'. The surface in question is the solid form of a mathematical relationship in three dimensions (x, y, z) which is the graphical expression of progressively more complex functions – quadratic, cubic and so forth (Figure 10.6). The degree of approximation to a selected surface can be tested and divergences from the surface (or 'residuals') can be calculated.

Figure 9.13 Rectangular and circular graphs
- Each rectangle is divided according to the percentage area in each land-use classification. It would have been possible to make each rectangle proportional to the total area of the country by varying the widths.
- Circular graphs divided according to percentages of various distributions, areas or occurrences. 360° is equal to 100 per cent and the circle is divided accordingly.

10

Map analysis

In the previous chapter attention was paid to ways in which precise information may be extracted from various types of map. The map may be regarded as a representation or *model* of the landscape and is a major source of material for geographical studies as a data bank; it has, however, been prepared for some specific purpose and may well be selective or even biased in this representation.

Much modern work in geography and allied subjects is in the field of analysis – locational, spatial, regional and so on. Map analysis may provide a starting point for some of these investigations but it generally needs to be supplemented by other observations, reports and census data.

Geographical investigations take many forms but when using mapped information it is possible to recognize four methods of approach, which may form successive stages of study. Such methods, or stages, may be undertaken at a variety of levels of understanding and are:

The recognition or descriptive stage: In map studies this can be regarded as the stage at which the model is translated into the map-user's view of reality.

The measurement stage: While the previous stage is chiefly qualitative in character it is generally followed by the need for precision, i.e. a quantitative approach must be made, in which some of the techniques outlined in Chapter 9 will be utilized.

The relationship stage: At this point the geographer is concerned with the patterns, locations, interaction and possible interrelationships suggested by the topographic and other distributions shown on the map.

The explanatory stage: Finally, theories will be put forward in order to account for the patterns noted and described.

For discussion purposes, these four stages will be regarded as distinct, but it is often difficult to separate them, since the mental processes involved are all interconnected.

RECOGNITION

The translation of the map into the details of the earth's surface that it represents is the starting point of any map analysis or, indeed, of any map reading. The map can only adequately be translated, however, if it has properly and accurately been prepared, and only then if the limitations of the map (its scale and so forth) are understood by the map-user. Map information is conveyed to the user by means of symbols, words and colour or shading. These must either adequately be explained by means of the map legend or else be completely self-explanatory.

Information on maps is conveyed also by the patterns of lines, shading and symbols and this is the point at which *recognition* and *relationships* can become confused. Thus, a town will be recognized by the name lettering, combination of housing and road symbols and possibly additional symbols such as railway lines, industries and the like. Nevertheless the primary activity will be that of recognition and identification of the main features of the town, the shape of the built-up area, the presence or absence of functions such as roads, rail routes, airports, industries an so on. Similarly, other geographical features may be recognized by simple relative patterns; for example, contour lines and water lines will give information about drainage patterns, river basins and watersheds. There is thus always an element of interpretation or assessment of relationships in the simplest descriptive recognition of the physical or human landscape (Figure 10.1).

Figure 10.1 Some linear patterns
  a Contours and drainage network
  b Routeways and settlements in an urban area

MEASUREMENTS

In order to carry out more precise methods of geographical analysis it is necessary to go beyond the initial and qualitative descriptive stage to that of quantification. A number of measurements that can be made from maps have been described in Chapter 9 and are summarized in Table 10.1. Such measurements will be used in conjunction with the description of the mapped topography so that the map-user can build a precise and detailed picture. Measurements will include details of shapes and possibly also of rank. A distinction may also be made between the static features of the topography and the more dynamic features presented on certain thematic maps. Nevertheless, the vast majority of maps present only information that is correct at a particular moment in time; that is, they are *synoptic* in character.

Measurements made from maps will need to be recorded in some way; it is most likely that this will be in a tabular manner. The data will then be available for statistical manipulation and further computations. It is also likely that some mapped information will still need to be compared

Table 10.1   Some selected measurements in landform and human (or cultural) geography

Measurement type	Landform data	Cultural data
Single values (by direct reading or interpolation	Spot heights Absolute height Relative height Stream order	Population figures Production data
Linear values (by 'scaling' against map scale bar)	Stream lengths Coastline lengths Watershed lengths Profiles Water flows Currents	Distances, either direct or route distances between selected features, towns, etc. Movements and flows Vectors
Areal values (by planimeter, geometry or sampling)	Drainage basins Hypsometric data Landform units Rock outcrops, soil associations	Land use data Administrative regions Percentage and density data
Angular values (by direct measurement or calculation)	Slopes Aspect and orientation	Movements population communications products, trade, etc.

with other records for the same locality, that is, its most significant factor will be its position relative to other points. The map co-ordinate system of locating points is therefore most important and may be incorporated with other tabulated information. If a considerable amount of information is to be recorded against a host of carefully located points, the most convenient way of assembling this information will probably be with some form of 'digitizing' equipment (Figure 10.2). Additionally, a variety of complex information can often be presented in a tabular form as a *matrix* (Figure 10.3).

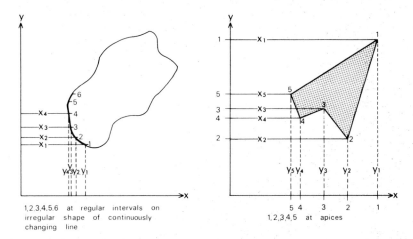

1,2,3,4,5,6 at regular intervals on irregular shape of continuously changing line

5 4 3 2 1
1,2,3,4,5 at apices

Figure 10.2 Digitizing: each change in direction of a line is recorded in terms or 'x' and 'y' co-ordinates that can be stored in a data bank, and the line subsequently reproduced at any conventional scale. Alternatively, change in direction can be recorded from a curved line by regular interval spacings at which 'x' and 'y' co-ordinates are recorded.

RELATIONSHIPS

Since the geographer's task is frequently that of describing and analysing the interrelationships that make up the landscape, many of the descriptive and quantifying tasks mentioned above are useful tools for the map interpretation of an area, large or small.

In very many cases the interpretation of the mapped data is a case of detecting *patterns* exhibited on the map. Such patterns may be in the form of lines (e.g. isolines or communciation linkages) or may be in patch or areal data (e.g. land use, buildings) as described in Chapter 3.

Simple patterns of either a linear or areal type are easy to establish and some possible examples are given in Figure 10.1. More complex patterns occur when a variety of information is presented and it is often desirable to simplify the map for purposes of analysis. This may be attempted in one of the following ways.

a Relative relief by 500 m squares, part of the Luxembourg Oesling, Troisvierges area (in metres)

a

30	30	25	35	25	15	15	60	40
20	25	15	20	15	25	30	55	25
20	15	5	30	40	25	40	50	50

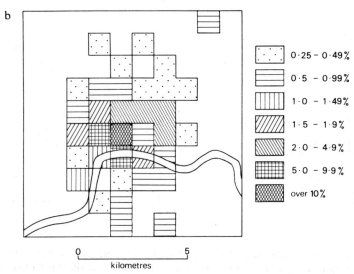

b

⠒⠒	0·25 – 0·49%
☰	0·5 – 0·99%
⫼	1·0 – 1·49%
⫽	1·5 – 1·9%
⧄	2·0 – 4·9%
▦	5·0 – 9·9%
▨	over 10%

0 ————— 5
kilometres

Figure 10.3
  a Section of table of data relating to grid squares, forming a matrix of information based on a map study in the Luxembourg Oesling
  b Cartographic representation of data collected on a grid-square basis ('Percentage of workers employed in printing', by courtesy of J. E. Martin)

1 *A 'sieve' method* Since the map (in particular, the topographical map) is composed of an overlay of several sets of information, usually in different colours, it is relatively easy to disentangle these sets by simple tracing or even by obtaining prints of the map of only the particular colour in which the student is interested, e.g. 'water and contour' prints showing only the brown and blue printing. These are sometimes easily available (Ordnance Survey and some European map series). Similarly, the total built-up area, distribution of particular land uses, patterns of communications and other factors can all be extracted and shown separately (see the various Land Utilisation memoirs for examples of this method) (Figure 10.4). Possible relationships can thus also be considered if two or more of these extracts are combined.

2 *Generalization methods* Frequently the detail of a particular distribution, and hence its pattern and relationships with other features, is concealed by the complexity of its occurrence, the number of isolated patches and so on. In this case some form of generalization may help to clarify the picture and probably the most well-known example of this is 'generalized contours'. The technique is applied in order to remove minor erosional landforms so that major forms can be distinguished (Figure 10.5). Such studies can be highly subjective in approach unless great care is taken, as an example, the application of some geometrical rules in order to distinguish

0 _____ 10
kilometres

Figure 10.4 Areal patterns of one distribution: grassland in the Fens from the Land Utilisation Survey memoir 'bevelling' or the drawing of 'generalization lines' tangential to concavities in the detail lines.

On the other hand, using point data extracted from the map together with the point co-ordinates, it is possible to recreate a 'surface' to which these data can refer, either factually as with erosion levels or hypothetically as with economic data, production and population statistics.

An alternative to this is the method of Trend Surface Analysis in which the point data is compared with mathematical surfaces of varying complexities and the degree to which these 'fit' can be seen and the displacements or 'residuals' measured (Figure 10.6).

Figure 10.5   Method of constructing generalized contours: the base map is part of the small scale relief map of Belgium (1/500,000) that shows the complexity of the contour pattern. Regional patterns of relief can be appreciated by generalizing the contours as shown.

3 *Sampling methods* In order to assess the occurrence of data, particularly land-use information or distributions that can be considered to cover substantial areas either individually or in aggregate on the map, it is useful to take a series of samples at various points on the map. Sampling techniques are varied but again normally make use of the grid-referencing system on the map together with a pattern of, for example, random sample points (Figure 9.7). Such a technique has been used with land-use data where it is required to ascertain percentages of each of a group of usages (Frolov and Maling 1969). Many other methods of studying relationships from mapped data are now available (see Harvey 1969 and Haggett and Chorley 1969) and stem from the realization that this information can be treated in mathematical ways that are not confined to those of the geometrical

Figure 10.6 Example of trend-surface maps: computer-calculated data from summit levels in the Sussex Weald compared with maps of theoretical or algebraic surfaces that approximately coincide with reality (after Thornes and Jones 1969). Surfaces shown by isolines representing (*top*) a linear function, (*middle*) a quadratic function and (*bottom*) a cubic function.

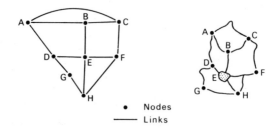

Figure 10.7   Nodes and links: the left-hand pattern is the simplified topological one, while that on the right shows the geographical reality: town with adjacent villages and roads linking the settlements. This pattern can be summarized in terms of twelve links and eight nodes (or vertices). Also, each point can be listed in terms of its number of linkages. Thus A, B, C, D, F and H all have three links, E has four and G only two.

distributions of the topography. Thus the techniques of topology, considering nodes and links, are of particular interest (Figure 10.7). The significance of map projections when considering distributions on small scale maps must also be remembered.

11

Mapping techniques

The exact methods by which a presentation of geographical information
will be made depends on:

The type and form of the information
The region or geographical area under consideration, and
The purpose of the presentation.

The range of possible methods is therefore very extensive, and the purpose of this chapter is to indicate possible steps in the cartographic presentation of data.

The raw data for the map will be gathered from a number of sources,
ranging from existing maps, statistical tables, written reports and guides
to air photographs and computer print-outs. These data will be capable
of being mapped as long as they refer to precise locations, which can be
either points, lines or areas.

As has been seen before (p. 7) a map can be regarded as being either
topographic or *thematic* in type. Most of the following discussion is concerned with thematic mapping methods and, more particularly, with the
preparation of cartograms, sketch maps and maps to be used as illustrations
or visual presentations in geographical studies. Technicalities and drawing
procedures will be considered in Chapter 12.

The first essential in any mapping operation is the preparation of the
framework; the mapping procedure follows one of the golden rules of surveying – 'work from the whole to the part'. The framework may take
several forms and may in fact be that of an existing topographic map either
to which data are to be added or from which some data are to be extracted,
or a combination of these two. In many cases a basic framework may have
to be constructed from existing maps and this will involve changes in
scale or in projection. Both of these changes may need to be computed
in order to draw the grid but the transference of detail will be by some visual
method (Figure 11.1). Detail is transferred by inspection from the square
or geometrical shape of the original to the corresponding square or shape

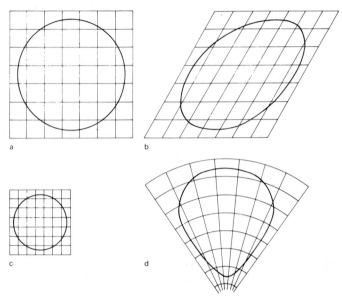

Figure 11.1 Variations in shape of a given outline as a result of transference to grids of different patterns: (a) and (c) show the use of grids at different scales in order to redraw outlines and detail at different scales. Methods (b) or (d) would be used in redrawing on a different projection. Method (c) provides an appropriate way to redraw at a different scale where photographic or optical methods are not available. In all cases the draughtsman must transfer line positions from old to new squares by visual estimation of intersections with grid lines, drawing more grid if necessary.

Figure 11.2 The optical method of changing scale: principle of the Grant projector, Plan-Variograph or Episcope. The drawing to be enlarged or reduced is placed on the carrier plate at the bottom of the diagram and the illuminated image is projected onto a drawing table. This may be by back-projection onto a sheet of tracing paper or film on a glass surface, but in the episcope the image is projected downwards onto a drawing surface. The lens position is adjustable and, in the Grant projector, so is the position of the carrier plate.

in the new grid. If necessary, additional shapes or squares will be drawn to subdivide those shown; diagonals and estimations of intersections with grid lines may also be used (Guest 1969, Monkhouse and Wilkinson 1971).

As mentioned above, the cartographer will frequently need to redraw part of a map at a different scale. This can be done by a visual method or by one of the following:

a Photographic processes, by direct photography of a map and its printing at an enlarged or reduced scale as required.

b Episcope methods, in which a tracing is made of the image at the required scale. Special equipment is available (e.g. Grant projector, Plan-Variograph) by which the original document is projected on a horizontal glass surface and the image enlarged or reduced as required. This can then be traced or photographed (Figure 11.2).

However, if no suitable existing map is available or if the base maps are on different map projections (Steers 1965) to the required map, the cartographer's first task will be the preparation of a framework (possibly by redrawing on a new grid – see above and Figure 11.1). If the information is made available by tabulated values referring to a rectangular grid this must be carefully drawn up on the drawing board or, if precision to fractions of a millimetre is required, by using equipment such as the *co-ordinatograph*.

The above information will be required both in the initial stages of mapping in order to assist the cartographer in the location of detail and also when the map-user is extracting information from the map. It is, however, background information and should be kept to the minimum so as not to provide too much 'noise' and obscure the main theme of the map. This can be done by using thinner lines or neutral colours. The locational data must be made of secondary importance. It can consist of one or more of the following:

1 A grid system – usually rectilinear
2 Outlines of natural features – coastlines, rivers and so forth
3 Outlines of land-use features – such as built-up areas and land parcels
4 Boundaries of administrative divisions
5 Significant routeways

and will also frequently include some place-names, grid-reference values and spot heights.

The primary thematic data of the map, i.e. the information message that the map is to portray, is presented in a bolder way than the background, secondary information. This can be achieved in the form of heavier line-work, the use of shadings and colour, and the addition of symbols (Figure 11.3). Explanatory lettering may be printed on the surface of the map but is more likely to form an extensive legend. Included in this category are the various graphs and diagrams that may be printed on the map and which can be viewed as appendices. These can include tables of data, short sections of text or graphs of statistics relating to specific places or areas shown on the map. Such insets are referred to as 'cartograms' but may themselves be diagrammatic map representations, which is another use of the same term.

Figure 11.3 Example of a thematic map in which information is portrayed by variations in style of line symbols ('Pipelines in the United States', *Philips Geographical Digest* 1970)

Crude oil pipelines
Products pipelines
Natural gas pipelines
Proposed gas pipeline
▲ Oilfields
△ Natural gasfields
● Major refinery centres

0   600
kilometres

It is therefore possible to distinguish the following problems in the making of thematic maps:

1 Considerations of map content: and together with this the map's purpose and aim must be considered.
2 Methods of presentation: under this heading are the factors of map scale, layout, legend and representation.
3 Map style: the form of linework, lettering types, symbols, colour shades, map shape and so on make up this section.

When decisions have been made on these three points the detailed map specification has been made.

Figure 11.4   Methods of data presentation

MAP CONTENT

This has already been discussed in general terms and it has been shown that two main categories of information are represented on thematic maps: the background information and the message of data that the map is to present. The aim of the map and the audience to which it is directed must also be considered at this stage.

Figure 11.5 Examples of some of the many symbols available for thematic map use: each symbol has been shown in two sizes in the ratio 3:1; they represent the following features:

Column i	Column ii	Column iii	Column iv
a electricity production	a iron and steel production	a wine production	a fortress
b atomic energy	b agricultural equipment	b brewery	b church or cathedral
c thermal electricity	c motor-car construction	c dairy produce	c important factory
d gas produc-tion	d aircraft construction	d panorama	d bridge or viaduct
e petrol refinery	e petrochemical industry	e viewpoint	e lighthouse
f coal	f rubber industry	f nature reserve	f vineyards
g iron	g brickworks	g conifers	g golf
h non-ferrous metals	h cotton and linen	h leafy trees	h tennis
j petroleum	j wool	j palm trees	j bus or coach station
k natural gas	k mills (grain and flour)	k prehistoric remains	k camping site

PRESENTATION

Most of the possible methods of data presentation have been considered in Chapters 2 and 3 and are summarized in Figure 11.4.

The graphical form in which the data is to be presented determines the amount of detail that can be shown. In the representation of statistics, geometrical shapes are generally used and the entire figure may well be a geometrical pattern – as in various forms of graph (Figure 9.13) or topological diagram (Figure 10.7). The superimposition of graphical shapes on a map outline leads to other diagrammatic forms (Figures 9.12 and 10.6). Examples of thematic maps are almost limitless (Figures 1.5, 3.2, 4.3, 10.3, 10.4, 10.5, 11.3, 12.4) but, in general, as scale increases a greater amount of detail can be shown.

The representation of detail on diagrams and maps varies with the type of information to be shown as well as with scale. At the largest scale it is usual to show all information in plan form together with explanatory names and/or abbreviations where necessary (e.g. official 1/1,250 or 1/2,500 plans). With decrease in scale outlines become generalized but names and abbreviations will still be shown as far as possible (e.g. 1/25,000) until all information is symbolized (as at the topographic map scales of 1/50,000 or 1/63,360). Thematic maps are sometimes prepared at large scales and will largely conform with these remarks, but the majority of distribution maps in which a shading represents quality and/or quantity of presence are at smaller scales (e.g. 1/250,000 or smaller). Choropleth and isopleth methods characterize such maps although these techniques may be used at all map scales.

MAP STYLE

Although the two previous factors will to some extent control cartographic details of style, it is necessary to make decisions on the following points:

1 Symbols
2 Linework widths
3 Colours and/or shading.

Before a final decision can be made on some of these points the scales of both the final map and that of the original, or fair drawing, must be decided. Normally the map will be drawn at a linear scale of one and a half, two or even two and a half that of the final map. All symbols, widths and lettering sizes must be chosen with the appropriate reduction factor in mind (Figure 12.5). This drawing will be reduced photographically to the correct reproduction size.

Map symbols (Figure 11.5) will usually conform to conventional usage and can be either qualitative or quantitative. In the latter case, considerable

Figure 11.6 Some different lettering styles and the resulting changes in general appearance of part of an atlas map

    a Mixture of *Albertus* and *Joanna* italic type: country names are set in Univers Medium. The Joanna italic has narrow serifs and no variations in 'weight' of lettering are used on this map. Differences are emphasized by changes of point size or of typographical style: thus towns in different Joanna italic, regions in Albertus.

b Type on this map is Univers throughout. Country-names are in Extra Bold (stippled) and a wide range of weight in the Univers type is used, from the very light river names to the extra bold of the larger towns.

c All lettering is in Univers type. Four weights of upright letters are used for town names and italic for physical features, rivers and spot heights.

d This map uses serif-style lettering (Times Bold) and also some condensed versions of Grotesque and Placard Bold.

care must be taken in their drawing, as the exact size of the symbol will have a precise meaning. Pre-printed symbols or a stencil can sometimes be used to maintain uniformity, but it is frequently necessary to construct symbols geometrically (Monkhouse and Wilkinson 1971, Hodgkiss 1970).

Table 11.1   Colour usage for different phenomena on maps

1 BLACK for details of settlement and ground-plan lines, point information, lettering and topographic features such as rock outcrops; also for boundaries
2 Strong BLUE for drainage network
3 Reddish BROWN for contours and also possible for rock outcrops (Grey can also be used)
4 Strong GREEN for vegetation, tree and woodland symbols
5 GREY for shading or hatching lines of built-up areas, buildings
6 YELLOW for any distributional tones required or additional boundary lines (as zone alongside lines)
7 Dark BROWN for additional detail
8 Strong RED for communications network
9 Light BLUE for areas of water, seas, lakes
10 Light RED available for higher relief shading
11 Light GREY for relief shading

The first three can be considered as the major colours of the linework on the topographic map.

These colours will show through those of the next group (4, 5 and 6) which serve to complement 1, 2 and 3. The remainder are supplementary colours.

Note that 1, 2, 3, 4, 7, 8 are strong colours, to be used for lines and symbols in the main.
5, 6, 9, 10 and 11 are light colours, to be used chiefly for shadings but possible for linework.
6, 9 and 10 can be used together and in various strengths to produce a range of shadings.

Blue and blue mixtures are generally regarded as *cold* colours; use these therefore in arctic and alpine areas, e.g. icefields, higher relief.

Red and red mixtures are generally regarded as *warm* colours.

Widths of linework will vary on the map as between lines of background information and those of greater importance; grid lines are usually kept to a minimum; and use will be made of various line styles (Figures 11.3 and 10.1), especially if the map is in a single colour.

The use of colour leads to specific problems, notably in the preparation of the map originals. Colour printing of maps is usually carried out by a 'separation' process in which a separate printing plate is prepared for each colour. These separate drawings must be made on stable material so that they exactly agree in size and shape. Also, 'register marks' must be placed on each separate drawing to assist in positioning over one another

(Figure 12.4). When colour tones or shadings are involved, care must be taken to ensure that appropriate gaps are left in other colours so that 'overprinting' does not occur; this may be desired, however, in some cases, depending on the colours involved.

The choice of colours is always a vexed one. Considerations of conventional usage, the map type and generalities of appearance and aesthetics are all involved (Keates 1962). However, a number of widely accepted conventions can be recognized and are summarized in Table 11.1. In addition, printing problems will make it necessary for any cartography involving colour maps to be a co-operative matter between map designer, cartographer and printer.

For uniformity of shading on any map it may be necessary to use pre-printed tones or symbols or stencils (Chapter 12), often causing some limitation in the range of shadings available to the map designer. If a series of maps are under consideration then additional care is needed in the choice of both colours and shadings, to preserve uniformity and to allow for the greatest possible range in data.

The numerous lettering styles and type faces available today provide the map-designer with considerable choice (Figure 11.6). Uniformity in lettering may be achieved by using stencils or by applying pre-printed letters and numbers (Figures 12.7 and 12.8). This is desirable in maps that form part of a series (for example, one sheet of a number covering adjoining regions) or a sequence (that is, several maps showing different facets of the same area) as well as those that are to be reproduced in one volume. Care must be taken to allow for any reduction factor between fair-drawing and printing. If a different reduction factor is used this will change the values of shading as well as of heights of lettering. Clarity and uniformity will make for easier reference and reduce the 'noise' factor. Usage on some topographic maps has led to the following conventions, which are *not*, however, universal:

Upright lettering for town and place-names.
Italic lettering for natural features, seas, rivers, etc. (or for areas and regions as opposed to settlements).
Gothic or Old English lettering for landscape features of antiquity.

12

Draughting techniques

This chapter contains a brief consideration of basic methods for the preparation of maps, diagrams and similar illustrative material for either publication or fair presentation. The cartographer must be a draughtsman and, although there are many aids to drawing, a minimum level of skill is essential. This can only be reached by practising individual methods and styles since map preparation involves the cartographer in acquiring an expertise in drawing, lettering and the layout and presentation of material. It is necessary therefore to consider both the 'tools of the trade' and their application to cartography.

DRAWING SURFACES

For best results it is necessary to use good materials. A considerable range of drawing materials are available from which the cartographic draughtsman will choose according to the type of job in hand. These materials include the following:

1 Paper and board (opaque)
2 Tracing paper and tracing linen (translucent)
3 Plastic film (opaque, translucent or transparent)
4 Enamelled surfaces, usually on metal, e.g. zinc (opaque).

Drawing paper is the least stable of these materials because it is liable to expand or contract with atmospheric changes, especially those of moisture. Tracing linen and tracing paper display the same problems but thin board, for example Bristol board, is more stable because of its thickness. It cannot be rolled or bent, however. Paper, board and cloth can be defined by their thickness characteristics, generally quoted in decimals of an inch or millimetre, but it is common to encounter descriptions in terms of:

a Paper type and quality, e.g. cartridge paper, Engineers, Kent made
b Drawing surface, e.g. smooth, glossy, art or hot pressed, rough

c Weight, for instance 80 lbs – referring to the final press weight in the manufacturing process

d Sheet size by terms such as Imperial and Demy; but these are currently being replaced by International sheet sizes (A0, A1, A2, etc.).

In the case of paper, these descriptive terms are usually combined in the total nomenclature, e.g.

'Antiquarian   Whatman's   Hot Pressed'

(size of sheet)   (paper quality)   (finish)

Plastic film comprises a range of drawing materials of different characteristics:

1 Clear film – transparent material
2 Opaque film, the surface of which can be cut or 'scribed' with clear lines
3 Opaque film, covered by surface capable of being 'peeled' to give clear areas
4 Translucent material.

A perfectly clear film is required for the preparation of display material for projection purposes, for example overhead projector transparencies. Translucent or tracing material is most commonly used as a general purpose plastic film in the preparation of maps when stability is important, and it is marketed by several commercial organizations under such trade names as 'Kodatrace', 'Ethulon', 'Permatrace'.

Surfaces that can be marked with lines by a fine steel or similar pen (Figure 12.1) ('scribed') provide stability and also allow for greater uniformity of linework than that given by applying ink. A commercial production of this is 'Scribecoat', usually red or orange in colour, but it may be specially prepared in other colours. Similar colours are used in the production of 'Peelcoat'. In this material an upper surface of plastic can be removed by first cutting round the desired shape and then removing by peeling (Figure 12.1). In this way areas to be shaded can be indicated in a negative manner or other parts of the map cleared to leave an opaque area where the shading or colouring is to appear (positive).

DRAWING EQUIPMENT

Most cartographic work is undertaken by drawing black or coloured outlines on a white background, and drawing equipment is therefore needed to apply these lines or colourings. For sketching purposes, preliminary layout work and so on, pencil lines will be used, while final linework will be in ink. Fair drawing on scribing material is carried out with a special

Figure 12.1 1 illustrates the use of Peelcoat drawing material. A is the opaque material, B a soluble adhesive and C the transparent film. S indicates the scribing point, C the dampened sponge and K a knife-point used to remove the Peelcoat; diagram 2 shows a simple outline before (a) and after (b) removal of opaque Peel-coat. The lower picture shows a scribing tool. The 'front' leg of the supporting head may be removed so that the operator can control the scribing pressure.

scribing pen that uses a sapphire tipped or steel point. Various designs of nib point allow for variations in width and style of line in both ink work and scribing.

Most linework will be aided by the use of guidelines, templates and stencils to ensure accurate geometry. Thus, straight lines are drawn against a machined straight edge, parallel lines drawn with a parallel ruler, perpendiculars and angles drawn by reference to a protractor or set-square. Many of these can be assisted by placing a sheet of accurately squared paper (graph paper) beneath the tracing paper (Figure 12.11) and/or by the use of a drawing table fitted with parallel guides, protractor and so forth (Figure 12.2).

Ink linework is the most commonly used method in cartographic draughtsmanship and the following pens and inks will probably be encountered:

1 Stencil pens, e.g. UNO, Leroy, Standardgraph, Rotring Varioscript
2 Line pens, e.g. Ruling pens, Rapidograph, Rotring Variant
3 Freehand or quill pens, including special lettering pens and nibs.
    Some examples of pens, compasses and their use form Figure 12.3.

All pens must be *filled* with ink and *not* dipped into an ink bottle. In this way ink will only be placed in the reservoir and not on the outside of the nib. It is necessary to ensure both a steady flow of ink and to reduce contact friction. The pen should be slightly tilted in the direction of movement. Pens attached to compasses are similarly tilted and in all cases movement should be made in a steady and deliberate manner.

Figure 12.2 A standard drawing table with fittings for parallel movement of a straight edge; the table can also be adjusted to a convenient working angle.

Parallel motion

Angle of board adjustment

a

b

c

Figure 12.3 Pens for ink linework, lettering with stencils and ink compasses
  a Ruling Pen
  b *Variant* drawing instrument (*Rotring*)
  c *Rapidomat* stand for *Rotring* pens (*Varioscript* and *Variant* types): this is a
    humidifier that ensures that pens are kept ready for use; the pen caps are
    arranged in front with the line width dimensions (in tenths of a millimetre)
    visible. Pen holder and attachment for stencil use can also be seen here.
  d Line thicknesses drawn by the various nibs in the *Rotring* range
  e Detail of the difference between the nib points: *Variant* point has a 'shoulder'
    to prevent ink from flowing under guide edges against which it may be used.
    Varioscript has no such shoulder as it is used vertically in stencils for lettering.
  f Using the *Variant* as a freehand drawing instrument for uniform linework
  g A drop compass for drawing very small circles
  h Ink compass in use showing reservoir and fitting

d

0.1 mm	
0.15 mm	
0.2 mm	
0.3 mm	
0.4 mm	
0.5 mm	
0.6 mm	
0.8 mm	
1.0 mm	
1.2 mm	

e

g

f

h

a

b

c

d

e   f

Figure 12.4 Stages in draughting a map: the diagrams on this and the facing page show a suggested series in which a two-colour map might be drawn.

   a Overall map size determined and border drawn (probably in pencil only), for inking over at stage (f). Symbols positioned. Register cross marks drawn outside map frame at all four corners.

   b Lettering added.

   c Linework (coast, frontier and edges of areas to be shaded) added.

   d Separate drawing made for second colour (pipelines and Delta area) with identical register marks in corners. Gap made where line crosses other detail.

   e Shading added to first colour drawing: note that, when using pre-printed dry transfer lettering together with pre-printed shading, care must be taken to prevent the latter adhering to the letters and removing them. It may be prudent in some cases to apply the shading before the lettering and cut 'windows' where the lettering will be placed. If line shading and stipple are hand drawn, this precaution will not apply.

   f Add legend, north point and title if required.

Map legend

1 Petrochemical industry	*2 Iron and steel industry*
3 Nuclear power electricity generation	*4 Pumped storage electricity generation*
5 South Holland oilfield	*6 Main industrial areas*
7 Land over 400 m	*8 Natural gas pipeline (in second colour)*

PROCEDURE (Figure 12.4)

A systematic approach is necessary in the drawing of a map and, although a possible series of stages is outlined below, the individual cartographer will proceed in an orderly way most suited to the types of work he is required to undertake.

1 Designing the map: In many cases the draft map will be prepared from some existing map and will probably be drawn in a variety of inks or pencil on tracing paper. The final 'fair drawing' is prepared by tracing over this on transparent or translucent material. In some cases the draft may be made in very light pencil on a good quality white surface, e.g. Bristol board, and this may be inked over with great care. The precise size of the draft map will depend on that of the final reproduction. These can be correctly related by simple geometry, as shown in Figure 12.5.

Figure 12.5 The relationship between the final reproduction size of the map and the size at which it is drawn are easily obtained by extending a diagonal as shown here. Also note the effect of reduction on lettering size. Other possible sizes of original or final print are shown by the broken lines.

2 Add single symbols: These may be qualitative, e.g. tourist sites, or quantitative, in which case the size of symbol gives an indication of town size, industrial production or other activity. Lettering and line-work must therefore conform to these symbols and thus the symbols will be drawn first.

3 Applying lettering (see below): Note that it may sometimes be prudent to leave this until last, especially if pre-printed and waxed lettering is used, as surplus wax can become a problem with subsequent line-work.

4 Line-in all solid and broken lines, making gaps where lettering has been placed and allowing for any break due to shading or symbolism still to be added. Work from the finest line to the heaviest if there

is doubt as to their distinction in the final map – lines can easily be thickened but the removal of ink is tedious. Complex linear symbols may be applied using a pre-printed adhesive tape produced under trade names such as *Chartpak, Normatape, Letratape* (Figure 3.1). Erasures can be made by various methods:

a India-rubber/abrasive erasers – hand-held or electrically operated
b Abrasive paste
c Sharp cutting edge, of knife or razor blade.

It is usually necessary to restore the drawing surface of the paper, film or other material. This can be done by burnishing the area of the erasure with a smooth tool of metal, hard plastic or ivory.
5 Apply tones, shadings, colour washes, etc. (see below).
6 Add margins, titles, legend, decorations and so on not included in basic lettering at stage 3.

Figure 12.6   Guidelines for positioning lettering: note that the writing follows the *general* line of roads and rivers rather than conforming to all the twists. Also note the position of heights, summit names and town names relative to the features.

LETTERING

Lettering can be added to maps and diagrams either by careful freehand work or by using one of several aids to lettering, such as stencils or pre-printed letters.

Freehand lettering may consist of single strokes or be 'built up'. Guidelines are essential to control both the height and the character of the letters, for example, angle of slope. Guidelines are often desirable in applying lettering by stencil or other methods and should always be used to control blocks of text or legends. A sheet of squared paper beneath the draft map being copied is of considerable assistance in lettering as well

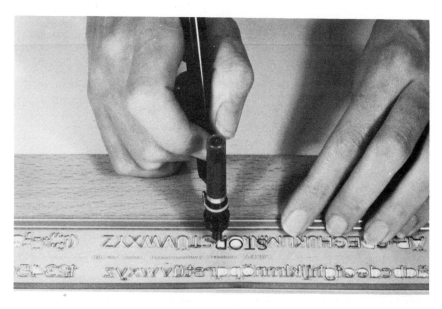

Figure 12.7 Using a stencil as an aid in lettering: the stencil is a transparent plastic guide held above the drawing surface by its raised edges. In order that the lettering is aligned correctly the stencil must be placed against a straight edge that is held firmly. The stencil pen, in this case a *Rotring Varioscript*, is held vertically by means of a special holder or adaptor. The stencil here being used is a *Standardgraph* that contains upper and lower case letters, numbers and some symbols

as shading but must be made 'square' with the map margins. However, the labelling of some features, such as rivers, hills, flow lines and route-ways, may need a curving word form, and guidelines are then drawn (Figure 12.6).

Stencils of a variety of character heights and styles are available and the types most commonly encountered are:

1 UNO, Standardgraph and others, where a plastic guide is held above the paper surface and controls the movement of a vertically held

(i)

(ii)

a

ABCDEFGHIJKLMNOPQRST UVWXYZÆŒÇØ abcdefghijklmnopqrstuvwxyzææç øß1234567890?!(%)&«»;:/,"

ABCDEFGHIJKLMNOPQRSTU VWXYZÆŒÇØ 1234567890 abcdefghijklmnopqrstuvwxyzæ œçøß?! (%)&«»;:/,"

b

ABCDEFGHIJKLMNOPQR STUVWXYZÆŒÇØ abcdefg hijklmnopqrstuvwxyzæœçøß 1234567890?!(%)&«»;:/.

ABCDEFGHIJKLMNOPQRST UVWXYZÆŒÇØ abcdefghijklmnopqrstuvwxyzææ çøß1234567890?!(%)&;:

c

ABCDEFGHIJKLMNO PQRSTUVWXYZÆŒ ÇØ abcdefghijklmnopq rstuvwxyzæœçøß 1234567890?!(%)&«»;:/+

𝔄𝔅ℭ𝔇𝔈𝔉𝔊𝔥𝔦𝔧𝔨𝔩𝔪𝔫𝔬𝔭𝔮𝔯𝔰 𝔗𝔘𝔙𝔚𝔛𝔜𝔷& 1234567890?!& abcdefghijklmnopqrstuvwxyz

d

ABCDEFGHIJKLMNOPQRS TUVWXYZÆŒÇØabcdefg hijklmnopqrstuvwxyzæœçø ß1234567890?!(%)[];:

ABCDEFGHIJKLMNOPQRS TUVWXYZÆŒÇØ ();: *abcdefghijklmnopqrstuvwxy zæœçø1234567890?!()*

Figure 12.8
  a Lettering typefaces, some of the many different alphabets available in a pre-printed form

Column i	*Column ii*
a Baskerville	a Futura
b Times Bold	b Times Bold Italic
c Clarendon Bold	c Engravers Old English
d Univers	d Univers Italic

ABCDEFGHIJKLMNO
abcdefghijklmnopq

ABCDEFGHIJKLMNOPQRST
abcdefhiklmnorstuvwxyz123

ABCDEFGHIJKLMNOPQR
abcdefhiklmnorstuvwxyz

ABCDEFGHIJKLMNOP
abcdefhiklmnopqrstuv

ABCDEFGHI
abcdefghijk

ABCDEFGHIJK
abcdefhijklmno

ABCDEFGHI
abcdefhijklm

ABCDE
abcdefg

ABCD
abcde

ABC
abcd

AB
abc

Figure 12.8
  b A range of print sizes in the Univers typeface: note that the size is not that
  of the actual letter but of the print body upon which it is placed. Left, top to
  bottom, 8, 10, 12, 14, 16, 20, 24 and 36 point. Right, 48, 60 and 72 point.

special pen (Figure 12.7). The pen can be obtained in various nib
thicknesses depending on the stencil in use.
2 LeRoy system, where a metal or plastic guide controls the movement
  of a specially equipped pen. The guide in this case is placed parallel
  to and below the lettering.

Pre-printed lettering has been used for many years in commercial
mapping offices and has now become widely used in drawing offices and
small organizations. A variety of lettering styles and sizes are available as

well as shades, symbols and colours. Two forms are generally available, both of which are wax adhesive.

 a A 'cut-out' variety, in which the individual letters or words are on thin transparent film that is made to adhere to the drawing by pressure. Examples are *Formatt* lettering sheets and the pre-set words produced by photolettering machines.

 b A 'transfer' variety in which the individual symbols are released by pressure from a backing sheet and are then applied directly to the drawing. Examples are the founts of lettering supplied in sheets under trade names such as *Letraset* (see Figure 12.8) and *Letter-Press*.

With the first category care must be taken in positioning the symbols after the removal of the backing material (Figure 12.9), while with the second it is necessary to be careful in the alignment of individual letters released from the fount sheet (Figure 12.10). With this type of pre-set symbol there is also a risk of mishap due to faulty adherence, but errors can be speedily rectified. Alignment is aided when working on transparent material by having a sheet of squared or lined paper beneath.

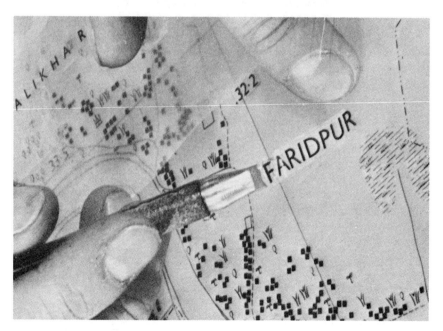

Figure 12.9 Positioning lettering of the pre-printed-on-film variety

SHADINGS AND AREAL SYMBOLS

Both freehand and mechanical methods can be used in applying shadings to maps. Freehand symbols are only possible if exact repetition can be obtained, as for example with lines or geometrical shapes. A useful aid to such freehand work on tracing materials, plastic films and similar transparent surfaces is a sheet of squared (graph) paper placed beneath the map draft. This will provide guidance for parallel lines, perpendiculars, spacing of dots and lines (Figure 12.11).

Symbols are frequently drawn with the aid of a stencil or, where large and geometrically simple, they may be constructed with compass and ruler. Alternatively, use can be made of pre-printed forms, again of both cut-out and/or dry transfer types (e.g. *Letratone, Formatt, Normatone*). Where separate symbols are used in combination to produce a complex shading, for instance for vegetation, some guidelines will be required – either as an underlay sheet or in light pencil.

AAAABBCCCDD
DEEEEEFFGGHH;
HIIIIJJKKLLLMM
NNNOOOPPQQR
RRSSSTTTTUUU
VVWWXXYYZZ
àààbbcccdddéé;
éééfgghhhiiiijjk
klllñññmmôôôop;
pqqrrrrssssttttu;
vvwwxxyyzz̄üüß
1̄1223344556677《
788900)(&?!£$¢%

Figure 12.10 Example of a 'fount' or portion of a sheet of *Rapitype* pre-printed dry transfer lettering: note that both upper and lower case letters are available, also numbers and some symbols.

COLOURS

In the preparation of a map that is not to be reproduced, a great variety of colours can be used and the range will only be restricted by the particular needs of the map, legibility and the skill of the cartographer. Water-colour paints are most likely to be used in the form of colour washes over larger areas, and colour inks over smaller areas.

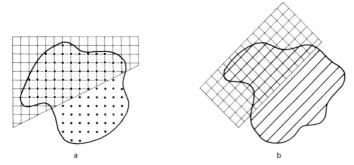

Figure 12.11 Using squared paper for guidelines: the squares can be used to aid in the location of lettering, the spacing of a dot shading pattern and for spacing a line shading pattern.

Preparation for the reproduction of a map in colour will involve the cartographer in working from a monochrome base map showing *all* the information and the drawing of separate 'maps' or 'plates' in black for each colour to be used. This base plate is referred to as the 'key' plate.

It is essential that all these separate plates fit together exactly and for this purpose 'registration marks' are drawn on each sheet (Figure 12.4). These will be lined up carefully at various stages in map reproduction and printing plate preparation; therefore the cartographer must work with equal care in drawing the individual colour outlines. Each outline must be prepared on similar drawing materials, and this must be stable in dimensions. In some cases it may be easier to prepare a 'negative' plate, that is, filling in those areas NOT to appear in the colour under consideration. An appropriate 'positive' will then be prepared photographically.

13
Map reproduction

The topic of *reprography* is of special importance to the cartographer because it is very likely that he will be preparing maps for publication or reproduction in some form. Also, when handling published maps, the map-user will find that a knowledge of the preparation stages will indicate limitations in the mapped data; for example, accurate measurements can only be taken from precise surveys printed on stable material.

Although the single map in its original form may sometimes be all that is required, multiple copies are often needed. The exact number will depend on the type of map and its degree of sophistication. One can distinguish, for example, between the clearly general-interest type of map represented by the tourist topographic map sheet (Ordnance Survey 1 inch map of Dartmoor, Figure 2.1c) and the specialist diagrams illustrating a paper, thesis or report (Figures 1.4, 3.2, 9.9 and 10.6). The former will be published in thousands of copies and can be regarded as 'long-run' printing jobs whilst the latter may require no more than a dozen copies to be made. Between these two extremes are the multitude of topographic and thematic maps and plans used both as separate publications and/or as illustrations in books and journals. The following categories can perhaps summarize the situation:

1 Maps produced in large quantities in which the techniques of conventional printing (single or multicolour) are economically practicable, e.g. small and medium scale topographic and thematic maps, atlases.
2 Limited publication runs of either separate map sheets or of illustrations in the text of books and journals. These are generally single colour (monochrome) but can be more complex. Normal printing processes are again desirable.
3 Small numbers of copies of individual maps and diagrams.

In the first two cases there is usually a stage of photographic reduction between the original fair-drawn map and the reproduction whereas in the third category 'true-to-scale' copies are the norm. Similarly, the vast majority of maps produced in small quantities are in one colour, frequently black on white.

A great variety of processes are available for the reproduction of maps and diagrams (Clare 1964, Hutchings 1970), including the following:

1 Letterpress or 'relief' methods of raised lines to which ink adheres and from which a print is made.
2 Etched or *intaglio* processes of engraving where the ink is carried in a channel. Recessed patterns also form the basis of the *photogravure* process.
3 Level printing surfaces of *lithography* where an inked image is transferred to the paper, generally by means of an intermediate 'blanket' roller.
4 Photographic processes involving chemical reactions to light to produce lines, including 'dye-line' and 'true-to-scale' copying methods.
5 Reproductions by stencil methods, either waxed stencils and ink or spirit duplicating. The stencils may be prepared by a photographic and/or electronic process.
6 Electrostatic copying methods of Xerography.

The first three techniques are used for 'long runs' of printing, that is, when a large number of copies is required, while the remainder are more likely to be used for smaller numbers, ranging from single copies by electrostatic or photographic methods to hundreds by the stencilling techniques. Where a 'wet' process is involved, for example development of photographic film or paper, distortions may occur and accuracy of measurement may be lost. Similarly, lines can become blurred in duplicating and electrostatic copying. Some of these various processes should be considered in slightly more detail.

LETTERPRESS PRINTING

Raised symbols, letters and designs have been used for printing since the days of woodcuts. The raised lines can either be placed on a level surface as in 'flat-bed' printing machines, or transformed into a curved printing surface that can be fastened to the rotary cylinder of a high-speed printing press.

The map to be printed must therefore be converted into a printing plate or line block; this is carried out by making a negative of the original reversed as in a mirror image, which is then exposed to the printing plate so that the resulting prints will be in the correct sense (Figure 13.1). Where shades or tones are to be reproduced a 'half-tone block' must be prepared. This is made by photographing the original through a 'screen' consisting of a grid of finely spaced lines. Tones are then broken into dots of varying size, depending on the intensity of the tone (Figure 13.2).

LITHOGRAPHIC PROCESSES

The original discovery of the basis of this technique of printing was in the late eighteenth century when it was found that an impression could be made from a greased outline drawn on a piece of limestone (*lithos* = stone). Present-day developments of this use the fact that two adjoining surfaces can be printing and non-printing areas if one is made capable of retaining ink and the other receptive to water, greasy printing ink and water being mutually exclusive. The process thus initially depends also on dampening of the paper or wetting of the plate.

Figure 13.1   Printing cylinder and impression produced from the mirror image of the printing plate

The lithographic printing plate is usually of aluminium which is chemically sensitized so that a contact print can be made upon it from the negative of the original map. The print lines are prepared with ink as stated in the next paragraph, and an impression made from these is then transferred to the paper, either directly or, more usually, by way of an intermediate 'blanket' roller (Figure 13.3). The actual printing process is thus not from the actual plate but is 'offset'; the inked plate will therefore be a positive.

The actual preparation of the printing plates involves their coating with a light-sensitive layer that will adhere to the image or printing areas after exposure to the original (on clear film or similar material). The non-printing areas will then be dissolved away. Plates are coated by a centrifugal method; the plate is placed in a 'whirler', a solution is poured over its face and an even covering is achieved by spinning the plate at a high speed together with controlled air and temperature conditions for drying.

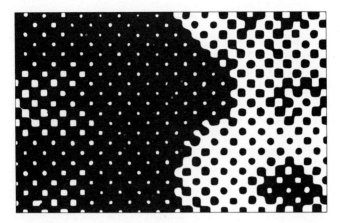

Figure 13.2 Magnified example of part of a 'half-tone' printing plate: the plate is prepared by photographing through a screen of fine lines that result in a pattern of squares. However, these tend to produce rounded outlines so, in detail, the pattern consists of a mass of dots of various sizes.

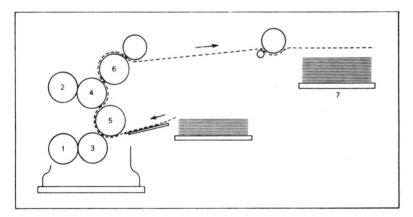

Figure 13.3 The main features of the offset lithographic printing press: the schematic diagram shows a two-colour press, and the direction of movement of paper through the press is arrowed.

1 Cylinder to which the printing plate for the first colour is attached
2 Ditto for second colour
3 Blanket cylinder transferring first colour image to 'impression cylinder'
4 Blanket cylinder for second colour
5 Impression cylinder
6 Transfer cylinder
7 Pile of printed sheets

ENGRAVING

Recessed or engraved lines on a printing plate retain ink while the smooth surface is wiped clean; the image of the lines or lettering is then transferred to the paper. This method was used in many early maps, the engraving being carried out by hand on copper plates (Figure 2.8). Early Ordnance Survey maps were prepared in this way, as were Hydrographic Department charts until comparatively recent times. The advantage as far as charts were concerned lay in the fact that amendments could quickly be carried out; the engraved line was relatively easily removed and new information added without destroying the entire sheet. Ink is also retained in depressions on the surface of the plate in photogravure methods of printing shades. The deeper the shade to be printed the deeper must the recessed cell be, so that a greater amount of ink can be held. These cells form a regular pattern over the printing cylinder and after this has been inked a scraper blade removes the surplus from the non-printing surface.

DYELINE REPRODUCTIONS (DIAZO PROCESSES)

When a quick copy of a single colour line map is required, and provided that the original is on a transparent or translucent material such as tracing paper or film, it is possible to make a 'contact' print that will be a positive by dyeline duplication. Sensitized paper is placed against the back of the original and while the two are fed by rollers through the machine they are exposed to a strong actinic light. The sensitized paper is then 'developed' by passing through an ammonia solution, and the line-work will then appear as reasonably dense lines on a pale background, generally with a greyish-blue tinge. The chief defect of this process is the lack of permanence of the print, which will tend to fade and discolour, as well as a certain loss of sharpness in line and detail work.

Appendix 1
National grid referencing and Ordnance Survey large scale map legends

The National Grid is printed on all Ordnance Survey Maps of Great Britain and is a system by which metric co-ordinates can be given to any point in these countries (see Chapter 1, and especially Figure 1.7). Depending on the map scale, references can be read to the nearest metre, 10 metres, kilometre and so on, and this is most easily done with the aid of a 'romer' and by measuring to the nearest grid lines west and south of the position under consideration, as follows:

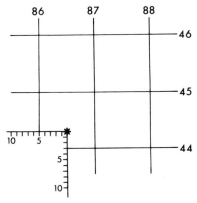

Appendix 1.1 Example of use of romer: the point indicated by the asterisk has co-ordinates of 865 East and 443 North. The '86' and '44' are obtained from the grid lines printed on the surface of the map (read from the margins of the map); the '5' and the '3' are read from the romer as shown.

To give a full grid reference at point P on the upper diagram on the next page:

1 Obtain from map margins the letters designating the 500 km and 100 km grid squares within which P lies.
2 Take the west edge of grid square in which P lies and read the figures printed opposite this line on north or south margin. Estimate tenths eastward.
3 Take the south edge of grid square in which P lies and read the figures printed opposite this line on east or west margin. Estimate tenths northward.

a 1/50,000

1" TO 1 MILE MAP

FULL GRID REFERENCE OF POINT P IS TO 264731

b 1/1,250

1/1,250 SCALE PLAN

FULL GRID REFERENCE OF POINT P IS TQ 26437315

Appendix 1.2   Some grid references on Ordnance Survey map series (not to scale)

Appendix 1.3   Conventional signs (pp. 139, 140)

REFERENCE CARD
NATIONAL PLANS AT SCALES OF
1:1250 or 50·688 inches to 1 Mile and 1:2500 or 25·344 inches to 1 Mile
(Information applicable to both scales except where otherwise stated)

SYMBOLS AND CONVENTIONAL SIGNS

.... Bracken *

.... Bush †

.... Coniferous Tree (Surveyed)

.... Coniferous Trees (Not Surveyed)

.... Coppice *

.... Deciduous Tree (Surveyed)

.... Deciduous Trees (Not Surveyed)

.... Furze †

.... Heath *

.... Marsh, Saltings

.... Orchard Tree

.... Osiers †

.... Osiers *

.... Reeds

.... Rough Grassland

.... Scrub *

.... Underwood †

† Used on plans surveyed or revised prior to 22·4·63.

* Used on plans surveyed or revised since 22·4·63.

Where space permits, tree and scrub symbols in woods are shown in groups of Four, Three, Two or One to indicate Close, Medium, Open or Scattered density of trees or scrub.

.... Antiquity (site of)

»»»···· Arrow showing direction of water flow

↑ B M .... Bench Mark (Normal)

↑ F B M .... Bench Mark (Fundamental)

⊗ .... Cave Entrance

▣ .... Electricity Pylon

ts .... Permanent Traverse Station

·rp .... Revision Point (Instrumentally fixed)

↑ rp .... Revision Point and Bench Mark (Coincident)

+ .... Surface Level

△ .... Triangulation Station

∫ .... Area Brace (1:2500 scale only)

Perimeter of built-up area with single acreage (1:2500 scale only)

Levelling Information: Altitudes of bench marks and surface levels are given in feet above the Newlyn Datum. Bench Mark Lists, which may contain later levelling information, are obtainable from Director General, Ordnance Survey.

Slopes				
Top		Sloping Masonry Top	Glazed Roof Building	Archways
Roofed Buildings	Culvert	Dunes	Sand	Shingle
Boulders	Flat Rock	Cliff Face	Active Quarry Chalk Pit or Clay Pit	Inactive Quarry Chalk Pit or Clay Pit

Walls, other than those of roofed buildings, which are less than 1 metre thick at 1:1250 scale or 1·4 metres thick at 1:2500 scale, are shown by a single line representing the centre of the wall.
Walls which are in excess of these widths are shown by two lines, each representing one face. For roofed buildings, the lines on the plan normally represent the outer face of the walls at ground level.

Thick Wall   Thin Walls   Represented by

for Building

1:1250 Scale Romer: To Measure Ten Metre Reference

Place arrowhead on Point with edges of scale parallel to grid lines on plan. Move the scale S and W until divisions coincide with grid lines. Read scale at these points, horizontally first and vertically second.

See reverse for 1:2500 scales and Romer; also for Boundary information and Abbreviations

9   8   7   6   5   4   3   2   1

BOUNDARY INFORMATION

England & Wales

─── ── ── .... County Boundary (Geographical)

· ─ · ─ · ─ · ....County & Civil Parish Boundary coterminous

· ─┼─ · ─┼─ · .... Administrative County or County Borough Boundary
L B Bdy

─⊖─ ─⊖─ ─⊖─ ....London Borough Boundary
M B Bdy   U D Bdy   R D Bdy

· · · · · · ·...........................County District Boundaries based on civil parish
R B Bdy

· · · · · · · ·.................................Rural Borough (Borough included in a Rural District)

England, Wales & Scotland

· · · · · · · · ....Civil Parish Boundary

Boro (or Burgh) Const   Co Const .... Parly & Ward Boundaries based on civil parish

Boro (or Burgh) Const & Ward Bdy .... Parly & Ward Boundaries not based on civil parish
Co Const Bdy

Scotland

(Not with parish)   (Coincident with parish)

─────── ─────── .... County Boundary (Geographical)
Co Cnl Bdy   Co Cnl Bdy

──┼── ──·──┼──·── ....County Council Boundary
Co of City Bdy   Co of City Bdy

──┼── ──┼── .... County of the City Boundary
Burgh Bdy   Burgh Bdy

· · · · · · · ....Burgh Boundary

─ ── ── ── · · · · ....District Council Boundary
Dist Bdy   Dist Bdy

Where the boundary of an Admin Co, Co Boro or Co of City is coincident
with that of a geographical county, the symbol for the latter is shown.

Examples of Boundary Mereings

⊙ .... Symbol marking point where
boundary mereing changes

· · · · Und · · · · · .... Undefined boundary

· · · Def · · · ....Original boundary feature destroyed or defaced

C B ....Centre of Bank	C O C S...Centre of Old Course of Stream		F W............................... Face of Wall
C C .... Centre of Canal, etc.	C C S............... Centre of Covered Stream		S R .... Side of River, etc.
C D .... Centre of Ditch, etc.	4ft R H ....4 feet from Root of Hedge		T B ....Top of Bank
C R.............Centre of Road, etc.	E K .... Edge of Kerb		Tk H........................ Track of Hedge
C·S .... Centre of Stream, etc.	F F .... Face of Fence		Tk S .... Track of Stream

ABBREVIATIONS IN FREQUENT USE

B H ....Beer House	N T .... National Trust
B P, B S ....Boundary Post, Boundary Stone	N T L........................... Normal Tidal Limit
Cn, C .... Capstan, Crane	P .... Pillar, Pole or Post
Chy ....Chimney	P C ....Public Convenience
D Fn .... Drinking Fountain	P C B .... Police Call Box
El P .... Electricity Pillar or Post	P T P ....Police Telephone Pillar
E T L ....Electricity Transmission Line	P O .... Post Office
F A P...........................Fire Alarm Pillar	P H ....Public House
F S ....Flagstaff	Pp .... Pump
F B ....Foot Bridge	S B, S Br .... Signal Box, Signal Bridge
G P ....Guide Post	S P, S L ....Signal Post, Signal Light
H .... Hydrant or Hydraulic	Spr .... Spring
L B .... Letter Box	S, S D .... Stone, Sundial
L C .... Level Crossing	Tk ....Tank or Track
L Twr .... Lighting Tower	T C B ....Telephone Call Box
L G .... Loading Gauge	T C P .... Telephone Call Post
Meml .... Memorial	Tr .... Trough
M P U .... Mail Pick-up	Wr Pt, Wr T. ......Water Point, Water Tap
M H .... Manhole	W B .... Weighbridge
M P .... Mile Post or Mooring Post	W ....·............................ Well
M S .... Mile Stone	Wd Pp .... Wind Pump

H or L W M M T ....High or Low Water Mark of Medium Tides (England & Wales)
H or L W M O S T .... High or Low Water Mark of Ordinary Spring Tides (Scotland)
M H or L W ....Mean High or Low Water (England and Wales) } On plans surveyed or
M H or L W S...........Mean High or Low Water Springs (Scotland) } revised after 12.11.64.

Made and Published by the Director General of the Ordnance Survey. Chessington, Surrey. 1967.

Romer for 1:2500 Scale

9 8 7 6 5 4 3 2 1

Appendix 2
Map projections

The earth we live on may best be thought of as a sphere. Strictly, it is an irregular and slightly flattened spheroid (ellipsoid) whose two axes differ by some 42 kilometres (12,714 kms for the minor axis between the poles and 12,756 kms for the major, equatorial, axis – see Appendix 2.1). However, for most cartographic purposes it is adequate to regard it as a sphere; this has the unfortunate property of being solid and without a truly flat or plane surface. Its surface is also completely continuous with no boundaries, and the contrast with a finite plane sheet of paper is therefore even greater.

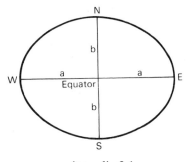

Appendix 2.1

These limitations mean that any representation of the earth will possess distortions when it is compared with reality unless it too is a three-dimensional solid, like a globe. Over small areas these distortions may be unimportant but they become increasingly vital as the map maker's scale decreases with its resultant extension of space. It is therefore very important to consider the map projection in all small scale mapwork.

It is possible to retain *some* of the geometrical features of the spherical earth on a flat sheet, but not *all* of them at any one time. Thus it is possible to make areas comparable between sphere and map, but at the expense of linear scale in some direction, even though it may be retained in another direction. Alternatively, angular properties, directions and shape may be

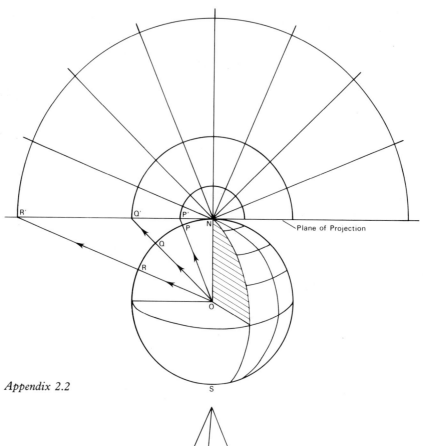

Appendix 2.2

Plane of Projection

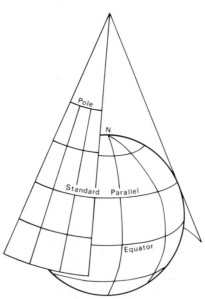

Appendix 2.3

retained at the expense of area. The technical terms for these two properties
are:

 EQUIVALENCE – preservation of areal relationships, on Equal Area
projections

ORTHOMORPHISM – preservation of shape so that at any single *point*
scale is identical in all directions. Scale will, however,
vary over the surface of the map.

The location system on the surface of the earth sphere is the graticule
formed by the network of lines of latitude and longitude (parallels and
meridians – Figure 1.6). These lines meet at right angles but this is not
necessarily the case with any particular map projection unless it is orthomorphic. Meridians are always Great Circles, like the Equator, but all
other circles are 'small circles' and thus less than the length of a full cir-
cumference of the earth. One way of examining possible map projections is
to categorize them according to the pattern of the graticule that results;
are these for instance straight lines, circular arcs or regular curves? Another
way is to consider the theoretical 'projection' involved. Three possible ways
of projecting the earth's graticule onto a flat sheet are shown in Appendix
2.2, 2.3 and 2.4). These are the bases of the Zenithal, Conical and Cylindrical groups of projection respectively.

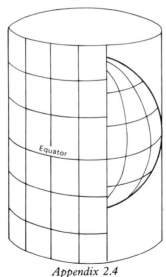

Appendix 2.4

The Zenithal projection assumes that one point of a plane touches the
sphere and it can be imagined that its properties are radial from this central
point; it (and its group) are also known as Azimuthal. In its simplest case
it is assumed that the point of tangency is one of the poles and the perspective aspect is indicated by projecting from the centre of the earth. Thus
point P is projected to P′, Q to Q′, etc. This particular Zenithal is the

GNOMONIC, which has the useful property that all Great Circles appear as straight lines; but it clearly has increasing distortion away from its central point. Other points of projection can be used (Appendix 2.5) e.g. from the antipodal point (South Pole in this case), which is the principle of the STEREOGRAPHIC projection, or by light rays from infinity as in the ORTHOGRAPHIC. Alternatively, instead of assuming a geometrical pattern of projection, distances from the centre may be calculated in order to preserve true scale distance from the centre (the EQUIDISTANT version) or

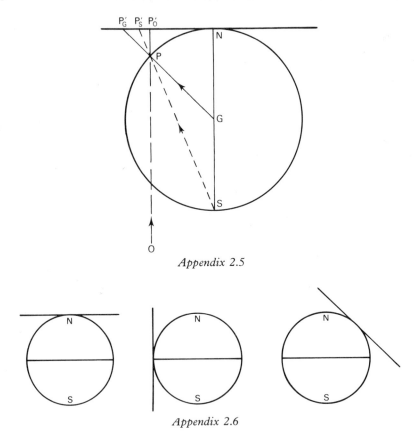

Appendix 2.5

Appendix 2.6

distances may be adjusted in order that areas are preserved (Zenithal EQUAL AREA). Again, the point at which the projection plane touches the sphere may be at the pole, the equator or anywhere (the Oblique case of this projection) and these are indicated in Appendix 2.6.

The Conical group of projections assume that the earth's graticule has been transferred to the surface of a cone that has then been 'unwrapped' (Appendix 2.7). Only part of the resulting shape will normally be used and adjustments may be made in the relative radii of the parallels in order to *having a common centre*

compensate for shape or areal distortions. The small circle of the latitude parallel at which the cone touches the sphere is known as the Standard Parallel and scale is correct along this line. The Polyconic version assumes that each parallel is 'standard' but these will no longer be drawn as concentric arcs although they may correctly be spaced on the central meridian. Derivatives of the Polyconic projection form the basis of more complex projections of the whole earth, for example the Hammer-Aitoff and Aitoff-Wagner projections.

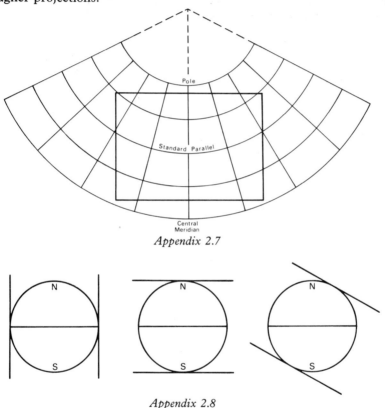

Pole

Standard Parallel

Central
Meridian

Appendix 2.7

Appendix 2.8

Similarly with the Cylindrical group of projections a solid shape that can be unwrapped (or *developed*) is intermediate between the sphere and the flat map. The length of the equator is thus the controlling factor for the length of the graticule of the whole earth but again different aspects of the projection are possible (Appendix 2.8 shows the normal, transverse and oblique cases). In order to preserve shape it would be necessary to enlarge scales at right angles to the central axis to compensate for the exaggeration parallel to it. Scale therefore becomes larger and larger away from this line but the resulting orthomorphic projection is well known – the Mercator. Alternatively, scale may progressively be decreased in order to preserve area, but this results in a very 'squashed' polar area in the normal case.

Various 'conventional' projections are used to obtain particular properties or to retain characteristics such as equivalence. Scale may often be
preserved along certain lines, for example along the parallels and the
central meridian (Bonnes and its derivative, the Sinusoidal) or equality of
area may be obtained by equating the surface area of the sphere to that of
a geometrical figure (e.g. the ellipse as in Mollweide's projection). Distortions may be minimized by making breaks in the projection shape – a
technique known as 'interrupting' the projection.

References

ARNBERGER, E. (1966) *Handbuch der thematischen Kartographie*. Vienna: Franz Deuticke.

BAGROW, L. (1964) *History of Cartography*. London: C. A. Watts.

BENNETT, H. A. *et al.* (1967) The land map and air chart at 1/250,000. *Cart. J.* 4, 89–95.

BIRCH, T. W. (1964) *Maps, Topographical and Statistical*. Oxford: Oxford University Press.

BLALOCK, H. M. (1972) *Social Statistics*, 2nd edition. New York: McGraw-Hill.

BOARD, C. (1967) Maps as Models in *Models in Geography*, edited by CHORLEY, R. J. and HAGGETT, P. London: Methuen.

BOARD, C. *et al.* (1970) in *Regional Studies* 4(3), October.

BOUGHTON, W. C. (1969) *Contour Plans by Computer*. Christchurch, N.Z.: Lincoln College of Agriculture.

BROWN, E. H. (1964) in *Guide to London Excursions*. London: IGU.

CARMICHAEL, L. D. (1964) Experiments in relief portrayal. *Cart. J.* 6, 18–20.

CARMICHAEL, L. D. (1969) The relief of map making. *Cart. J.* 6, 18–20.

CHORLEY, R. J. and HAGGETT, P. (eds.) (1967) *Models in Geography*. London: Methuen.

CHRISS, M. and HAYES, G. R. (1964) *An Introduction to Charts and their Use*. Glasgow: Brown, Son and Ferguson.

CLARE, W. G. (1964) Map reproduction. *Cart. J.* 1, 42–8.

CLARK, D. (1963) *Plane and Geodetic Surveying*, Vol. I, 5th edition. London: Constable.

CLAYTON, K. M. (1953) Denudation Chronology of the Middle Trent. *Trans. Inst. Brit. Geogr.* 19, 25–36.

COLE, J P. and KING, C. A. M (1968) *Quantitative Geography*. London: Wiley.

COLEMAN, A. (1965) Some technical and economic limitations of cartographic colour representation on Land Use maps. *Cart. J.* 2, 90–4.

CRONE, G. R. (1966) *Maps and their Makers*. London: Hutchinson.

DAVIDSON, Viscount (1938) *Final Report of the Departmental Committee on the Ordnance Survey*. London: HMSO.

DICKINSON, G. C. (1969) *Maps and Air Photographs*. London: Edward Arnold.

DICKINSON, G. C. (1973) *Statistical mapping and the presentation of statistics*, 2nd edition. London: Edward Arnold.

DIELLO, J. *et al.* (1969) The Development of an automated cartographic system. *Cart. J.* 6, 9–17.

DIXON, O. M. (1967) The selection of towns and other features on atlas maps of Nigeria. *Cart. J.* 4, 16–23.

DURY, G. H. (1960) *Map Interpretation.* London: Pitman.

ENVIRONMENT, DEPARTMENT OF (1970) *Atlas of the Environment.*

FROLOV, Y. S. and MALING, D. H. (1969) The accuracy of area measurement by point counting techniques. *Cart. J.* 6, 21–35.

GAITS, G. M. (1969) Thematic mapping by computer. *Cart. J.* 6, 50–68.

GALE, D. W. (1965) Register control in map production. *Cart. J.* 2, 68–74.

GEOGRAPHICAL MAGAZINE (October 1969) *Cartography for the 1970s.*

GOULD, P. and WHITE, R. (1974) *Mental Maps.* Harmondsworth: Penguin.

GUEST, A. (1969) *Advanced Practical Geography.* London: Heinemann.

HÄGERSTRAND, T. (1967) The computer and the geographer. *Trans. Inst. Brit. Geogr.* 42, 1–19.

HAGGETT, P. (1965) *Location Analysis in Human Geography.* London: Arnold.

HAGGETT, P. and CHORLEY, R. J. (1969) *Network Analysis in Geography.* London: Arnold.

HARLEY, J. B. (1975) *Ordnance Survey Maps – a descriptive manual.* Southampton.

HARVEY, D. (1969) *Explanation in Geography.* London: Arnold.

HINKS, A. R. (1947) *Maps and Survey,* 5th edition. Cambridge: Cambridge University Press.

HODGKISS, A. G. (1970) *Maps for Books and Theses.* Newton Abbot: David and Charles.

HOWARD, S. (1968) A cartographic data bank for Ordnance Survey maps. *Cart. J.* 5, 48–53.

HUNT, A. J. (1968) Problems of population mapping. *Trans. Inst. Brit. Geogr.* 43, 1–8.

HUTCHINGS, E. A. D. (1970) *A Survey of Printing Processes.* London: Heinemann.

IMHOF, E. (1965) *Kartographische Geländedarstellung.* Berlin and New York: Walter de Gruyter & Co. (English edition in press.)

IMHOF, E. (1966) *Schweizerischer Mittelschulatlas.* Zürich: Lehrmittelverlag des Kantons.

JOHANSSON, O. (1969) Further developments of producing photomaps as a basis for the economic map of Sweden. *Cart. J.* 6, 103–7.

JOLY, F. (1964) *Proposal for standardization of symbols on thematic maps.* (Paper presented to International Cartographic Association meeting, Edinburgh.) Paris.

KEATES, J. S. (1958) The use of type in cartography. *Surveying and Mapping* 18, 75–6.

KEATES, J. S. (1962) The perception of colour in cartography. *Proc. Edinburgh Cartographic Symposium.* Glasgow.

KERN, R. and RUSHTON, G. (1969) Mapit: a computer program for production of flow maps, dot maps and graduated symbol maps. *Cart. J.* 6, 131–7.

KILFORD, W. K. (1963) *Elementary Air Survey.* London: Pitman.

LOCK, C. B. (1969) *Modern Maps and Atlases.* London: Clive Bingley.

LYONS, H. G. (1914) Relief in cartography. *Geogr. J.* 43, 233–48 and 395–407.

MACKAY, J. R. (1953) The alternative choice in isopleth interpolation. *Prof. Geogr.* 5, 2–4.

MAGEE, G. A. (1968) The Admiralty chart: trends in content and design. *Cart. J.* 5, 28–33.

MALING, D. H. (1968) How long is a piece of string? *Cart. J.* 5, 147–56.

MCGRATH, G. (1965 and 1966) The representation of vegetation on topographic maps. *Cart. J.* 2, 87–9 and 3, 74–8.

MCKAY, C. J. (1967) The pattern map. *Cart. J.* 4, 114.

MEINE, K. H.(1966) Aviation cartography. *Cart. J.* 3, 31–40.

MERRIAM, M. (1965) The conversion of aerial photography to symbolized maps. *Cart. J.* 2, 9–14.

MINISTRY OF HOUSING & LOCAL GOVERNMENT (in progress since 1955) *The Desk Planning Atlas of England and Wales.* London.

MONKHOUSE, F. J. and WILKINSON, H. R. (1971) *Maps and Diagrams*, 3rd edition. London: Methuen.

MUEHRCKE, P. C. (1978) *Map Use – Reading, Analysis, and Interpretation.* Madison, Wisconsin.

ORDNANCE SURVEY (1951) *An Introduction to the Projection for Ordnance Survey Maps and the National Reference System.* London: HMSO.

ORDNANCE SURVEY (1968) *Parcel Numbers and Areas on 1/2500 Scale Plans.* London: HMSO.

ORDNANCE SURVEY (1969) *Its History, Organization and Work.* Southampton: HMSO.

ORDNANCE SURVEY, Map Catalogue, latest date, published annually.

OSBORNE, R. H. (1970) A Geographical Survey of Net Migration Movements in England and Wales, 1951–1961. *Revue Roumaine de Géologie, Géophysique et Géographie: Série de Géographie* 14(1), 123–35.

OVINGTON, J. J. (1962) An outline of map reproduction. *Cart.* 4, 150–5.

PANNAKOEK, A. J. (1968) Problems encountered in producing the National Atlas of the Netherlands. *Cart. J.* 5, 8–15.

PENROSE ANNUAL. London: Lund Humphries.

PETRIE, G. (1962) The automatic plotter and its consequences. *Proc. of the Edinburgh Cartographic Symposium.* Glasgow.

PETRIE, G. (1966) Numerically controlled methods of automatic plotting and draughting. *Cart. J.* 3, 60–73.

PHILIPS GEOGRAPHICAL DIGEST (published annually). London: Philips.

PIKET, G. (1968) The Water Control Map of the Netherlands. *Cart. J.* 5, 16–27.

RAISZ, E. (1962) *Principles of Cartography.* New York: McGraw-Hill.

ROBERTSON, J. C. (1967) The Symap Programme for computer mapping. *Cart. J.* 4, 108–13.

ROBINSON, A. H. (1965) The future of the international map. *Cart. J.* 2, 23–6.

ROBINSON, A. H., SALE, R. D. and MORRISON, J. (1978) *Elements of Cartography*, 4th edition. New York: Wiley.

ROSING, K. E. (1969) Computer graphics. *Area* 1, 2–8.

ROSING, K. E. and WOOD, P. A. (1971) *Character of a Conurbation.* Sevenoaks: Hodder & Stoughton Educational.

ROYAL SOCIETY (1966) *Glossary of Technical Terms in Cartography.* London: British Association of Cartography.

SCOTT, L. (1969) Early experience of the photomapping technique. *Cart. J.* 6, 108–11.

STEERS, J. A. (1965) *An Introduction to the Study of Map Projections.* London: University of London Press.

STORRIE, M. and JACKSON, C. I. (1967) A comparison of some methods of mapping census data of the British Isles. *Cart. J.* 4, 38–43.

TARRANT, J. R. (1970) Comments on the use of trend surface analysis in the study of erosion surfaces. *Trans. Inst. Brit. Geogr.* 51, 221–2.

THORNES, J. B. and JONES, D. K. C. (1969) Regional and local components in the physiography of the Sussex Weald. *Area* 2, 13–21.

THROWER, N. J. W. (1972) *Maps and Man.* Englewood Cliffs, N.J.: Prentice-Hall.

TOBLER, W. R. (1965) Automation in the preparation of thematic maps. *Cart. J.* 2, 32–8.

TOPFER, F. and PILLEWIZER, W. (1966) The principles of selection, a means of Cartographic generalization. *Cart. J.* 3, 10–16.

WINCH, K. L. (ed.) (1976) *International maps and atlases in print*, 2nd edition. London: Bowker.

WOOD, M. (1968) Visual perception and map design. *Cart. J.* 5, 54–64.

YOELI, P. (1967) Mechanization in analytical hill-shading. *Cart. J.* 4, 82–8.

ZUYLEN, L. VAN (1969) Production of photomaps. *Cart. J.* 6, 92–102.

The serious student of cartography should note the existence of the following publications, which form outlets for research work in the subject in the countries concerned:

Great Britain: *The Cartographic Journal* (British Cartographic Society) and the *Bulletin* of the Society of University Cartographers
USA: *The American Cartographer*
Canada: *Canadian Cartography*
Australia: *The Cartographer*

These are all produced by the various cartographic societies in the respective countries. Additionally, the following cover a somewhat wider field:

International Yearbook of Cartography – contains articles in English, French and German
Imago Mundi – an international publication in the field of historical cartography.

Index